COAST LINES

COAST LINES

MARK MONMONIER

HOW MAPMAKERS FRAME THE WORLD AND CHART ENVIRONMENTAL CHANGE

THE UNIVERSITY OF CHICAGO PRESS • CHICAGO AND LONDON

Mark Monmonier is distinguished professor of geography at Syracuse University's Maxwell School of Citizenship and Public Affairs.

The University of Chicago Press, Chicago 60637
The University of Chicago Press, Ltd., London

Printed in the United States of America

17 16 15 14 13 12 11 10 09 08 1 2 3 4 5

ISBN-13: 978-0-226-53403-9 (cloth)
ISBN-10: 0-226-53403-0 (cloth)

Library of Congress Cataloging-in-Publication Data

Monmonier, Mark S.
Coast lines: how mapmakers frame the world and chart environmental change / Mark Monmonier.
p. cm.
Includes bibliographical references and index.
ISBN-13: 978-0-226-53403-9 (cloth: alk. paper)
ISBN-10: 0-226-53403-0 (cloth: alk. paper)
1. Coastal mapping. 2. Environmental monitoring. I. Title.
GA108.7.M65 2008
912'.1946—dc22
2007035429

♾ The paper used in this publication meets the minimum requirements of the American National Standard for Information Sciences—Permanence of Paper for Printed Library Materials, ANSI Z39.48-1992.

To Reds Wolman,
with gratitude for Physiography of the United States—
an inspiring introduction to geography and how to teach it.

CONTENTS

PREFACE AND ACKNOWLEDGMENTS

I got the idea for this book while brainstorming chapter outlines for new projects. What's the most important feature on a map, I wondered, and what might uncover new insights and fascinating stories. Exactly when or why the light bulb lit up I don't recall, but the environmental hazards course I teach seems a likely stimulus. Whatever the inspiration, I became intrigued with the coastline as a unifying theme for exploring the map's evolving role in navigation, coastal zone management, and international politics. I'd written before about the hidden meaning of map symbols—necessarily generalized and easily manipulated, as I point out in *How to Lie with Maps*—and the prospect of focusing on a single type of map feature presented an exciting challenge. The project also meshed well with my work as editor of volume 6 of the *History of Cartography*, which deals with the revolutionary technologies that have spawned many new map genres in the twentieth century.

The split plural form of *coastlines* in the main title reflects the book's central theme: the emergence of four distinct cartographic coastlines. The first of these is the high-water line, visible from the sea and captured in diverse and often puzzling ways by European explorers with only a vague sense of longitude. The first coastline helps frame road maps, topographic maps, and wall maps as well as their smaller-scale counterparts in atlases and travel guides. It's portrayed with comparative precision on navigation charts, which also include the low-water line as a second coastline marking the tidal datum or sounding datum, to which depth measurements are referenced. Tied to the accurate measurement and reliable prediction of the tides, this second coastline is largely a nineteenth-century invention, the result of bigger vessels, systematic hydrographic surveys, improved tidal data, and a heightened concern for maritime safety. More recent are the shorelines associated with the respective inundations of severe coastal storms and sea level rise. Tailored to plausible assumptions about severe storms or climate change, these third and fourth coastlines warn of the short- and long-term dangers of imprudent coastal development.

Whether mapped from land, sea, air, or space, the coastline challenges mapmakers in diverse ways. Cartographic coastlines that evolved during the Age of Discovery reflect secrecy, ignorance, and the persistence of error as well as incremental improvements in technologies for estimating longitude and fixing position. In making meaningful, uncluttered maps at various scales, coastal cartographers must deal with a highly irregular land-water boundary perturbed by tides and coastal storms and further complicated by rocks, wrecks, and shoals. Captivating anecdotes include the frustrating effort to expunge fictitious islands from nautical charts, the tricky measurement of a coastline's length, and the contentious notions of beachfront property and public access.

My focus on the coastline was especially helpful in condensing the rich history of nautical charting into a concise treatment of the distorted continents and fictitious islands resulting from incomplete exploration and erroneous longitudes. As this admittedly eclectic selection of cartographic curiosities attests, *Coast Lines* is neither a conventional history of nautical charting (even at a broad brush level) nor an institutional history of charting agencies like the U.S. Coast and Geodetic Survey. An obvious concentration on twentieth-century developments reflects the importance of overhead imagery and electronic technology. Because my examples are drawn largely from the United States, the book is not fully representative of mapping practices worldwide. And with a focus on the societal impacts of cartographic coastlines, which often reflect assumptions and beliefs as much as science and technology, *Coast Lines* is not a textbook on marine mapping.

Rather than rehash material well covered in existing histories of pre-twentieth-century marine charting and coastal surveying, I move along quickly to examine the impacts of aerial imagery and electronic technology, the struggle to cover the entire world with moderately detailed standardized maps, and the role of coastal cartography in defining and contesting offshore boundaries. Separate chapters explore the third and fourth coastlines, and the penultimate chapter celebrates the diversity of graphic symbols required for customized portraits of the land-water boundary.

Although consciously writing for a general audience, I cannot ignore key parts of a story that gets a bit technical at times. In combining relevant bits of maritime history with the history of technology, *Coast Lines* integrates developments in mapping with issues in coastal hazards, coastal zone management, global warming, homeland security, the Law of the Sea, and marine navigation. Technical advances include electronic navigation, geographic information systems (GIS), global positioning systems (GPS), and satellite remote sensing. Readers can't understand the second coastline without knowing something about tides, for instance, and they can't appreciate flood-insurance maps and evacuation plans without grasping the underlying principles of storm-surge models. Because pictures are often more effective than words in making complex concepts accessible, I include a few explanatory diagrams among the book's more numerous maps and map excerpts. While shrewd readers can gloss over the more technical paragraphs—I do this myself at times—they should be able to appreciate underlying needs, general strategies, and significant effects.

My research depended heavily on helpful scientists at several federal agencies. Eric Anderson, senior geographer at the U.S. Geological Survey, put me in touch with Cindy Fowler, GIS program manager at the NOAA (National Oceanic and Atmospheric Administration) Coastal Services Center, in Charleston, South Carolina. Cindy and her colleague Mike Rink provided much useful information, including historic T-sheet images. I am likewise indebted to Skip Theberge and John Cloud, in the NOAA Central Library, at NOAA headquarters, in Silver Spring, Maryland. Skip read the finished manuscript and his collection of digital historic images was a tremendous resource, which I exploited numerous times throughout the book. John, who is working on a history of NOAA, provided valuable pointers to important but obscure documents. Also at NOAA headquarters, physical oceanographer Todd Ehret arranged a close-up look at Tide Predicting Machine No. 2, Curtis Loy filled me in on the techniques of boundary measurement, and Doug Graham provided details on past and current applications of photogrammetry and remote sensing. Jim Titus, who heads up the Sea Level Rise Program at the U.S. Environmental Protection Agency, was exceptionally generous in sharing insights, data, and publications. Also important were discussions of coastal hazards and emergency management with my colleagues on the National Research Council panel that prepared the recent report *Successful Response Starts with a Map: Improving Geospatial Support for Disaster Management*.

I also benefited from the gracious and conscientious assistance of numerous librarians and archivists in and around Washington at the Library of Congress, the NOAA Central Library, the National Library of Medicine, and the U.S. Geological Survey as well as at Cornell University, the New York State Library, in Albany, and the New York Public Library. At Syracuse University, John Olson, associate librarian for Maps and Government Documents, and Elizabeth Wallace, associate librarian for Science and Technology Services, were especially supportive. For help in identifying the enigmatic fifth island at Five Islands, Maine, I am grateful to Lynne Jones, president of the Georgetown (Maine) Historical Society. Cécile LePage and Diane Rioux of Tourism and Parks New Brunswick, Canada, and Mike Bonin of the Canadian Coast Guard were helpful in selecting and describing a photo dramatizing the tides in the Bay of Fundy.

Stephen Leatherman, director of the International Hurricane Research Center at Florida International University, reviewed a draft of the manuscript and provided a wealth of information about numerical modeling and its role in coastal risk mapping. I appreciate the amiable assistance of Mark Crowell at the Federal Emergency Management Agency and Keqi Zhang at the International Hurricane Research Center. I am also grateful to Keith Clarke, in the geography department at the University of California, Santa Barbara, and Ron Grim, at the Boston Public Library, for helpful comments on the finished manuscript.

Syracuse University's Maxwell School of Citizenship and Public Affairs awarded a semester's leave that let me visit collections and plug holes in my understanding of nautical charting, international boundaries, and coastal dynamics. Brian von Knoblauch and Stan Ziemba of Maxwell's Instruction and Computing Technology group were there when I needed them. In the geography department, staff cartographer Joe Stoll offered pointers on making Freehand and Photoshop do what I wanted, and research assistants Karen Culcasi, David Call, and Peter Yurkosky helped jump-start the project with a preliminary working bibliography.

Because working with clever, creative people who have helped me in the past is always a delight, I am especially grateful (again) to Christie Henry, my editor; Mike Brehm, who handled design; and Stephanie Hlywak, in charge of promotion. I also appreciate the vigilant eye and useful suggestions of Mara Naselli, my manuscript editor, and Mark Reschke, her colleague, who filled in admirably when Mara was on maternity leave. And as with previous projects, my wife Marge provided food, encouragement, and solitude when I needed them.

ONE

DEPICTION AND MEASUREMENT

On a top-ten list of mapped features, the coastline is a shoo-in for first place. Because the sea provides food, transportation, and recreation, the shoreline is at once a boundary, an attraction, a source of livelihood, and a hazard. On maps it reproduces the distinctive shapes of Africa and Cape Cod, and in textbooks and science magazines it reconstructs continental drift and dramatizes the changing climate's rising seas. A challenge to mariners and marine scientists, the coastline is raised and lowered twice daily by tides and occasionally realigned by storms, which famously shorten the shelf life of nautical charts, on which its representation demands careful measurement and prudent compromise.

Not all maps have a coastline. Whether a map has one often depends on its scale, which cartographers define as the ratio of map distance to ground distance. At a scale of 1:1 a map would be as large as the territory represented, and hopelessly cumbersome. At 1:87 it would shrink a minuscule part of the world to the detail of an HO-gauge model railroad—smaller than reality but large enough to show sidewalks and sewer lines. Most maps have much smaller scales. At 1:24,000 the U.S. Geological Survey (USGS) topographic maps favored by hikers and earth scientists compress an acre of land into a tiny square barely a tenth of an inch on a side. Even so, mapmakers consider these maps large scale, while a 1:200,000,000 world map a mere eight inches across is unquestionably small scale.

Small-scale world maps need coastlines to delineate continents, but less than half of all large-scale topographic maps include a stretch of seacoast.

Scale also affects the intricacy of cartographic coastlines, illustrated in this chapter by several maps for Five Islands, Maine, a tiny fishing village on Sheepscot Bay, thirty miles east of Portland. Marge and I rented a cottage there one summer when our daughter was young, and because we've been back a few times, I know the area well enough to appreciate the 1:24,000 USGS topographic map (fig. 1.1), which shows a settlement of roughly eighty buildings, mostly summer homes. (I enlarged this black-and-white excerpt to clarify symbols better differentiated in color on the original.) One of the tiny black rectangles represents our vacation rental, and another pinpoints the local "lobster pound" down near the dock, where the Thibodeau family sold the best take-out steamers on the Maine coast. Despite the simplified shoreline, I can recognize the narrow, rocky beach where Jo collected shells and flat pebbles. As a scale-model memory prompt, maps make excellent souvenirs.

Fine print at the bottom of the map sheet indicates that USGS mapmakers based their seaward coverage largely on the hydrographic measurements and shoreline features in figure 1.2, extracted from a nautical chart published at 1:15,000 by the U.S. Coast and Geodetic Survey, now the National Ocean Service (NOS) and part of the National Oceanic and Atmospheric Administration (NOAA). Intended for maritime users, charts present a comparatively sketchy view of the land, usually only for a narrow belt along the coast. In this example, topographic detail stops at a coast road a quarter mile inland—no point in cluttering the map with symbols of limited use to sailors. By contrast, the seaward portion of the chart is rich in soundings (depth measurements), bathymetric contours (lines of equal depth), and other symbols that warn of shallow water where a ship could run aground.

On nautical charts the coastline is more than just a line. Pictograms identify rocky areas submerged at high tide, and asterisks locate isolated rocks alternately covered and exposed. South of Malden Island the label "Foul" identifies an area where individual hazards remain uncharted and anchorage is risky. The USGS map

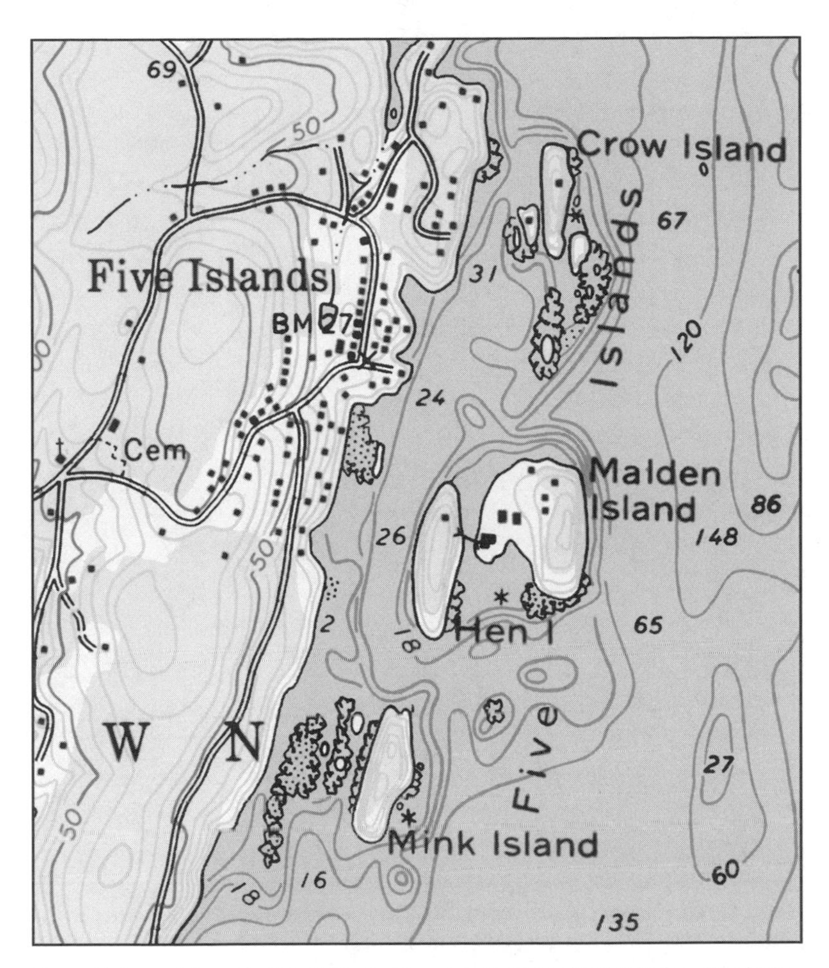

Figure 1.1. Excerpt from the U.S. Geological Survey's Boothbay Harbor, Maine, 7.5-minute topographic map, published in 1979 at 1:24,000 but enlarged to 1:14,400 for clarity. Area shown is 0.91 miles (1.46 km) wide.

(fig. 1.1) includes some of the soundings and most of the depth contours, but the NOS chart highlights unsafe water with a darker blue and ropes off dangerous zones with dotted lines. And while the topographic map names all four larger islands (Crow, Hen, Malden, and Mink), the chart identifies only the two that can be labeled without disrupting hydrographic detail. Trade-offs abound in mapmaking, and on a nautical chart hidden dangers trump feature names.

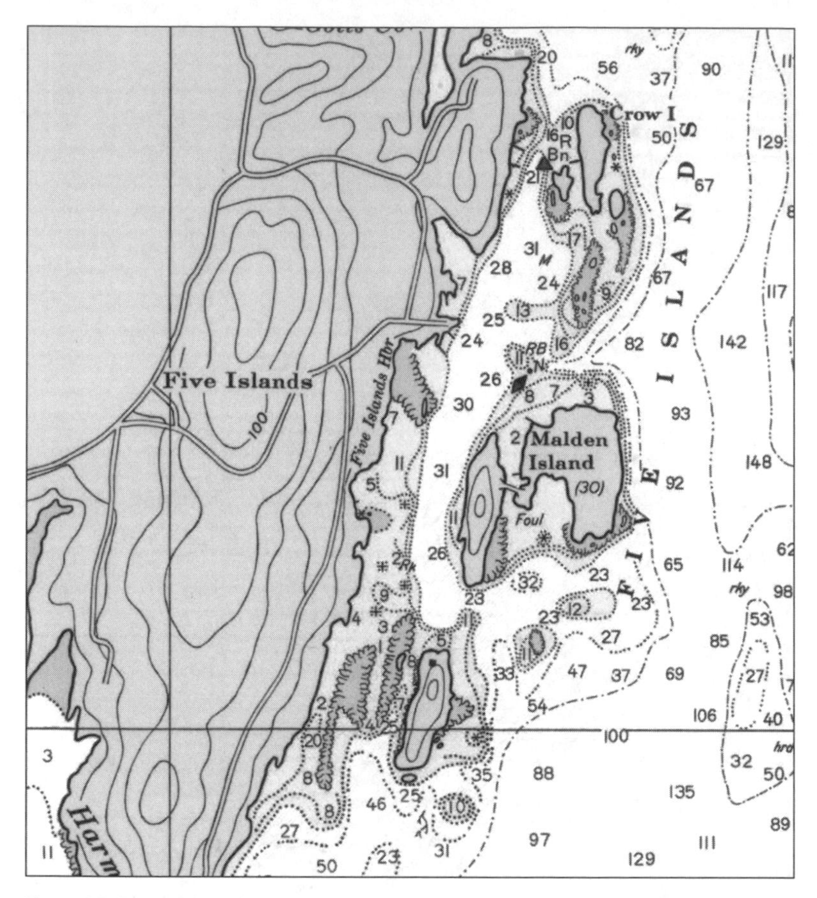

Figure 1.2. Excerpt from the U.S. Coast and Geodetic Survey Chart 238, Kennebec and Sheepscot River Entrances, published in 1970 at 1:15,000. Area shown is 0.95 miles (1.52 km) wide.

To illustrate more fully the effect of scale, I juxtaposed the shorelines from these two large-scale maps with corresponding delineations from topographic maps at 1:100,000 and 1:250,000. As figure 1.3 shows, the 1:15,000 nautical chart accommodates not only a more detailed shoreline than its 1:24,000 topographic counterpart but a few additional tiny islands as well. By contrast, at scales of 1:100,000 and 1:250,000, islets disappear and the shoreline becomes smoother and more rounded. At the smallest scale (fig. 1.3, *lower right*) the markedly distorted footprints of the four surviving islands are more

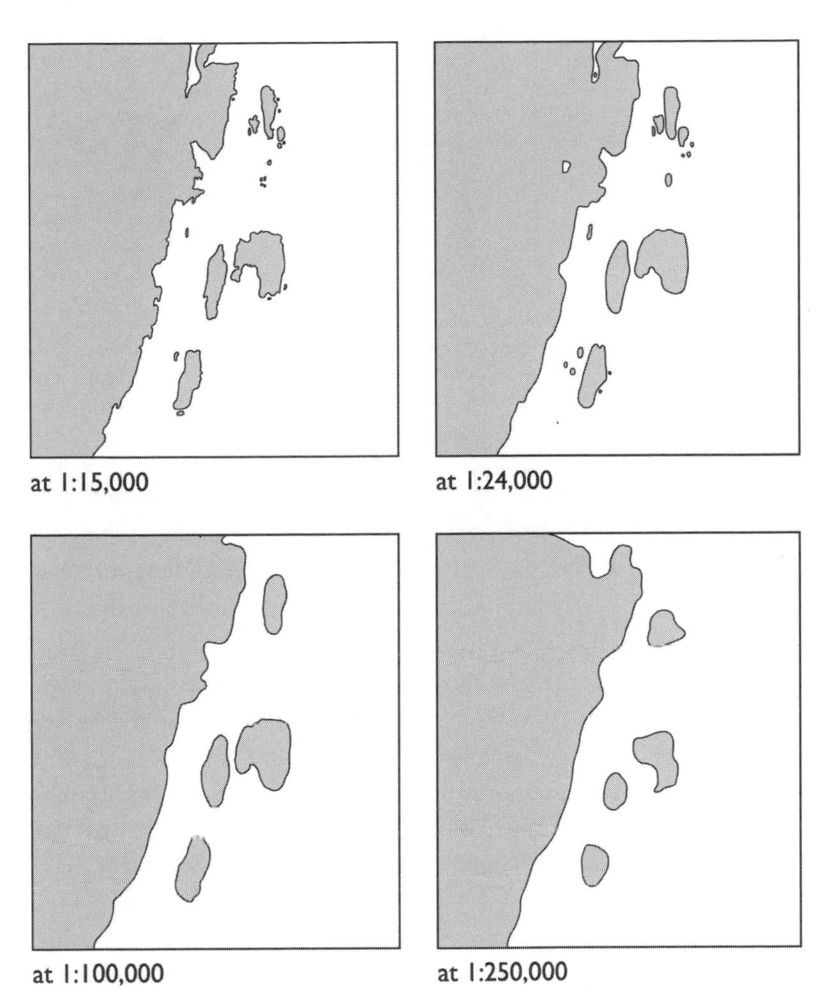

Figure 1.3. Corresponding coastlines, reduced or enlarged to a common scale, from maps originally published at 1:15,000, 1:24,000, 1:100,000, and 1:250,000. Excerpts cover the area shown in figure 1.1.

nearly similar in size, while Hen and Malden Islands, which are close enough to be connected by a footbridge, were pushed farther apart for clarity.

Displacement of features is common to coastal maps because a line symbol $\frac{1}{50}$-inch (0.51 millimeter) wide at 1:250,000 represents a band 417 feet across on the ground—nearly seventeen times its theoretical width at 1:15,000. This leaves the mapmaker with three

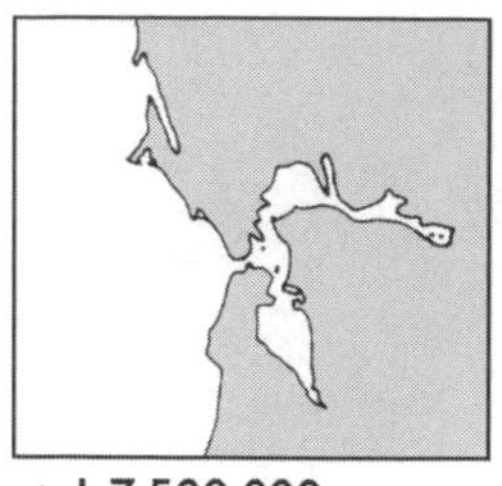

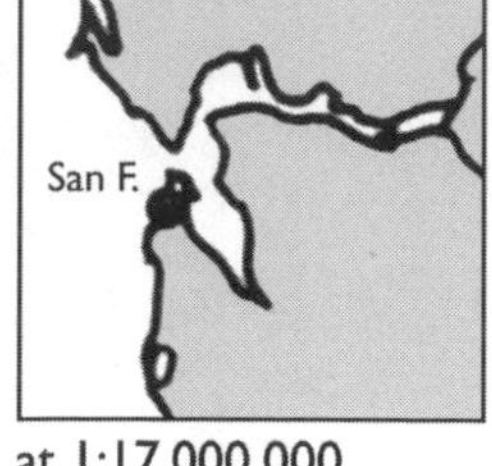

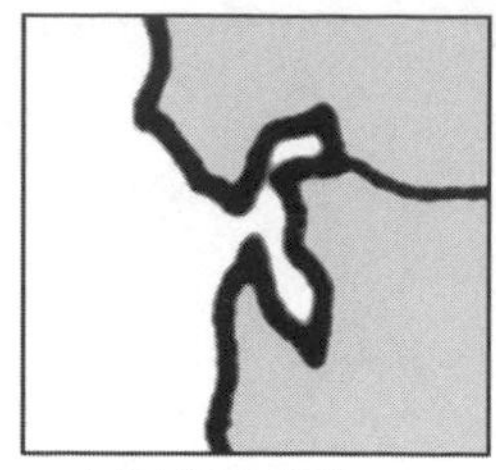

Figure 1.4. Generalization of the California coast near San Francisco. Map excerpts at 1:7,500,000, 1:17,000,000, and 1:34,000,000 were enlarged to the same scale from transparent overlays included with the *National Atlas of the United States*, published in 1970.

choices: amalgamate the two islands, shift them apart, or drop the smaller one. The 1:15,000 chart's more generous canvas accommodates numerous twists, turns, protuberances, and indentations, while the 1:250,000 topographic map must shoehorn an area 278 times larger into the same size frame.

Features that survive at smaller scales are often exaggerated. For example, the narrow inlet at the top of the 1:15,000 and 1:24,000 depictions becomes a distinctly wider incursion at 1:250,000. Why the map includes this feature is not obvious. Perhaps the cartographer who compiled the 1:100,000 topographic map thought the indentation dispensable, while the compiler of the 1:250,000 map, aware that appearances matter on reference maps, enlarged it to make the coastline look right. The smaller the scale, the greater the cartographic license.

Verisimilitude is particularly important for Maine's numerous long, narrow embayments, aligned north-south and similar in origin to the deep, steep-sided, glacially eroded fjords of the Norwegian coast. Small-scale maps typically retain at least a few of these drowned valleys, suitably widened and evenly spaced, as classic examples of a generalization process cartographic theorists call exaggeration.

A similar rationale governs portrayal of the California coastline near San Francisco. As excerpts from the *National Atlas of the United States* show (fig. 1.4), small-scale maps avoid overlapping symbols by widening the inlet linking the Pacific Ocean with San Pablo Bay (to the north) and San Francisco Bay (to the south). At 1:7,500,000 and 1:17,000,000 the mapmaker saw fit to include the broad Sacramento–San Joaquin River Delta (to the east) and the prominent coastline near Point Reyes National Seashore (to the northwest), where narrow, linear Tomales Bay follows the well-known San Andreas Fault. This lesser feature drops out at 1:34,000,000 because a line symbol $\frac{1}{100}$ inch wide would consume as much space as a corridor more than five miles across on the ground. At still smaller scales San Francisco's bays either drop out altogether or amalgamate into a tiny, barely noticeable dimple on an otherwise smooth coastline.

Because preferred routes and navigation hazards demand greater detail, chart makers employ a wider variety of scales than their topographic kinfolk, who rely on a comparatively uniform treatment to map a larger area for a more diverse clientele. As geographically tailored navigation instruments, harbor charts at scales between 1:5,000 and 1:20,000 promote safe access to ports, while in-shore navigation charts between 1:40,000 and 1:80,000 describe channels, hazards, and navigation aids for approach channels and longer stretches of coast, and off-shore navigation charts at 1:100,000 and smaller are sufficiently detailed for crossing oceans, seas, and gulfs.

Large-scale charts generally portray smaller, more focused areas than small-scale charts, and the area shown—what I call the map's geographic scope—is often chosen to encompass an entire estuary, port, or channel, usually identified in the chart's name. Because multiple map sheets can be inconvenient, if not confusing, nautical charts are often larger than topographic maps. For example, Chart 238, my source for figure 1.2, measures thirty-three by forty-four inches, whereas the Boothbay Harbor, Maine map sheet from which I extracted figure 1.1 is a mere twenty-two by twenty-seven inches.

Unlike their topographic counterparts, nautical charts vary in shape as well as size in order to fit meaningful chunks of the coastline onto a single sheet of paper. As figure 1.5 illustrates for coastal Maine, ease of use requires occasional overlap as well as supplementary, larger-scale charts for ports, rivers, and other areas where navigation is comparatively complex. By contrast, USGS topographic maps partition the country into nonoverlapping, one-size-fits-all quadrangles defined by an arbitrary grid of evenly spaced meridians

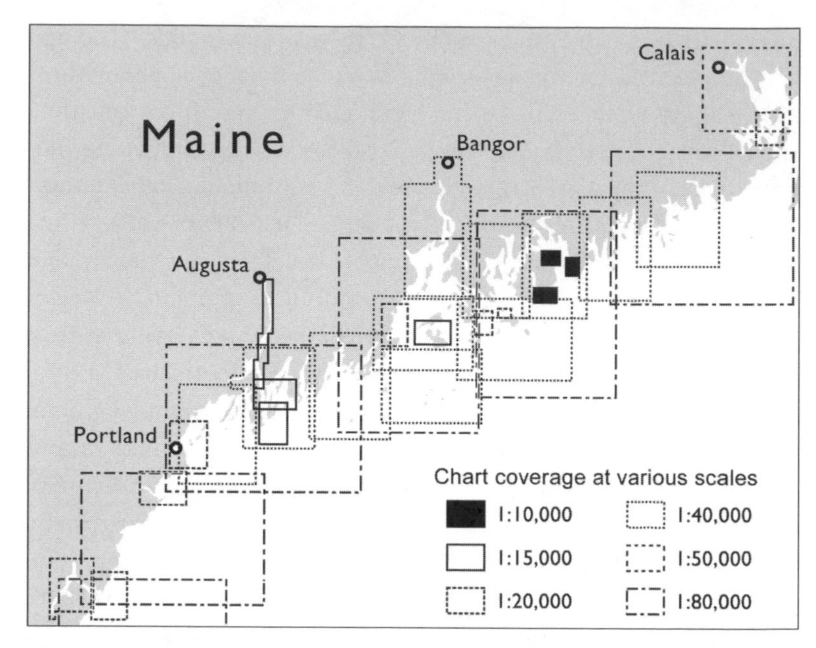

Figure 1.5. Geographic scope of nautical charts covering the Maine coast. Several charts include insets (not shown) covering smaller areas at larger scales (1:10,000 or 1:15,000).

and parallels, which all too often divide the area of interest among two or more map sheets. What's more, rigorous adherence to a uniform scale not only shortchanges areas for which fuller detail might be useful but also wastes space on largely featureless territory. An extreme example is the Rozel Point SW, Utah 1:24,000 quadrangle map, which portrays a rectangular chunk of the Great Salt Lake in solid blue. I taped a copy to my office door as an amusing example of cartographic obsession.

Map scale is not the only index of cartographic detail. A new measure emerged in the 1970s, when geographers began to study land cover with satellite imagery. Called "resolution" because it reflects an electronic sensor's ability to detect, or "resolve," features on the ground, this new index refers to the pixels (picture elements) of a digital image captured from a satellite or aircraft. Because pixels are usually square, resolution is expressed as the ground distance, in meters, along one side. Although an image can be displayed at any map scale, its resolution limits the level of detail, as illustrated in

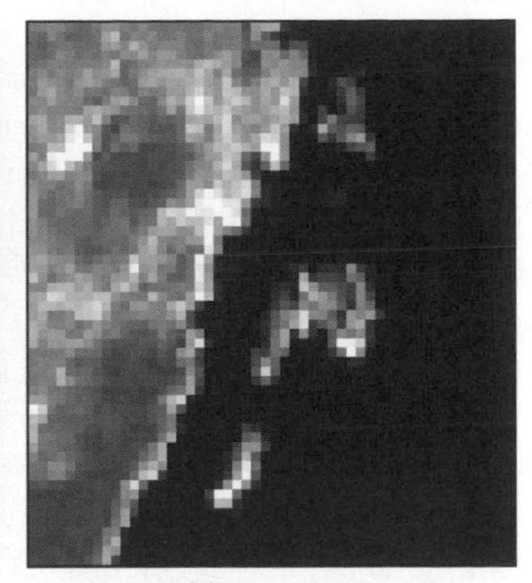

resolution: 1 meter

resolution: 28.5 meters

Figure 1.6. Examples of images with resolutions of 1 meter (*left*) and 28.5 meters (*right*). Excerpt on the left is from a Digital Orthophotoquad image from the Maine Office of GIS. Excerpt on the right is from an enhanced version of an orthorectified Thematic Mapper band 5 image from Landsat 5, obtained from the University of Maryland's NASA-sponsored Global Land Cover Facility.

figure 1.6 by two overhead views of Five Islands. With a resolution of one meter, the aerial image on the left not only shows roads and clearings but also provides a sharper picture of the tidal flats than the satellite image to its right. Open water stands out in dark gray or solid black on the 28.5-meter (93.5-foot) satellite picture, while its coarse pixel grid blurs the shoreline like the fuzzed-out faces of confidential informants in a television documentary.[1] At a much smaller scale, pixels would not be noticeable and the shoreline would seem well defined.

Though the two images in figure 1.6 resemble photographs, they are merely graphic renderings of numbers recorded for grid cells arranged in rows and columns. The left-hand image looks like an air photo because its grid is not obvious and its pixels portray the intensity of reflected sunlight largely as the human eye would see it. Look closely and you can see the wake of a speedboat cutting across parallel waves into deeper water. By contrast, the satellite image to its right records the relative intensity of infrared light reflected from the sea, the beach, clearings, and treetops. Invisible to human eyes, infrared distinctions would go unnoticed without infrared film or electronic sensors. Because water absorbs infrared radiation, pixels wholly offshore look black while light tones pinpoint healthy vegetation or bare sand, both highly efficient in reflecting infrared light.

A similar framework is used for digital elevation models (DEMs), which record elevations estimated for the centers of grid cells, typically thirty meters on a side. Like satellite images, gridded elevation data represent the shoreline as a zigzag boundary between onshore cells with a positive elevation and offshore cells, with elevations set to zero. When the resolution is a meter or less, or the map scale relatively small, a DEM provides a useful portrait of the coastline. But when the map scale is large relative to resolution, the display will look blocky, as in the right half of figure 1.6.

Map scale poses a serious limitation on measurements like the "95,000 miles of shoreline" that the Coast Guard cites when appealing for money or patience. A NexisLexis query of major newspapers and wire services found this estimate in over fifty news articles, 93 percent in post-9/11 stories on national security. Although 95,000 miles of coastline seems an intimidating responsibility, this figure is only one of several available estimates, which vary widely depending upon which maps one measures.

I traced the 95,000-mile figure to Aaron Shalowitz's *Shore and Sea Boundaries,* a massive compendium of legal and technical details published in 1964.[2] Shalowitz was the Coast and Geodetic Survey's boundary guru in the postwar years, and even though he never mentioned the 95,000-mile total, readers can add together the 88,633 miles of "tidal shoreline (detailed)" tabulated by state, the 1,338 miles reported (but not totaled) in a table called "Areas Over Which the United States Exercises Sovereignty," and the 4,678 miles cited in a single paragraph labeled "Shoreline Along the Great Lakes." The resulting sum, 94,649 miles, rounds to 95,000.

The U.S. Coast and Geodetic Survey measured the country's 88,633-mile maritime boundary in 1939–40 using a "recording measure"—a small hand-held device that counts the revolutions of a tiny wheel run carefully along a line—"on the largest scale maps then available."[3] Although the technique is not otherwise described, it was common practice in college map-analysis classes to make three measurements and take their average. If one measurement seemed an obvious blunder, it was thrown out and another taken. Nowadays we measure a curved line once, by summing up the individual lengths of its many short, straight segments, captured directly from aerial imagery or by carefully tracing delineations on existing maps and charts. The data and technology are available for revising the 1939–40 estimates, but replication is not a high priority.[4] And because several federal agencies use the data to allocate funds among the states, re-measurement could be highly political were the numbers to change radically.

When NOAA rises to the task, as it surely will, noteworthy revisions seem likely insofar current charts are generally more detailed than their predecessors. And even though electronic calculation does not sidestep arbitrary decisions about how far inland to follow an estuary, modern technology allows experimentation with multiple definitions, impractical for prewar coastal cartographers intimidated less by political fallout than the tedium of manual measurement and the need for consistency. As Shalowitz described the process, the "shoreline of bays, sounds, and other bodies of water was included to the head of tidewater, or to a point where such waters narrowed to a width of 100 feet."[5] At that point, the technician added a hundred feet to the running total and resumed measuring on the other side. A further qualification addressed long, narrow tidal rivers, which could inflate the results. "Both shores of a stream were measured if over 200 yards wide, but streams between 30 yards (100 feet) and 200 yards in width were measured as a single line through the middle of the stream."[6] However straightforward, these round-number thresholds, in feet and yards, are blatantly arbitrary. If Thomas Jefferson had persuaded Congress, a century and a half earlier, to replace traditional English weights and measures with a decimal system, the criteria and the resulting measured lengths would surely be different.[7]

For completeness Shalowitz supplemented his listing of each state's detailed tidal shoreline with much earlier estimates, published in 1915, for its "general coastline" and "general tidal shoreline."[8] The former was based on charts "as near the scale of 1:1,200,000 as possible" and a "unit of measure" of 30 minutes of latitude (approximately 34.5 statute miles)—tied to the earth's circumference but nonetheless arbitrary. A technician "walking" a pair of dividers along the mapped coastline added another 34.5 miles to his running sum at every step. By ignoring narrow estuaries and most coastal details, the 30-minute increment stepped off a coastline only 12,383 statute miles long. By contrast, a general shoreline based on 1:400,000 charts (or their 1:200,000 counterparts, where available) and a three-mile step yielded a countrywide total of 28,889 statute miles.

Among the states, step length clearly made a difference for Maryland, which registered only 31 miles of general coastline in contrast to 452 miles of general tidal shoreline, principally on Chesapeake Bay—far shorter, though, than its 3,190 miles of "detailed" shoreline. In theory a coastline could register a much greater length if measurement were based on an infinitesimally small step applied to a shoreline weaving in and out between grains of sand.

Which of these three sets of estimates is most appropriate depends upon its intended use. For example, the 95,000-mile figure is definitely more relevant to coastal-zone management than the 12,383 miles of general coastline. Even so, the need for comparable detail can make comparatively crude measurements defensible, as in a 1965 State Department study based on worldwide map coverage at 1:1,000,000 and ignoring estuaries less than ten miles across.[9] In that assessment, the United States accounted for only 11,650 miles of the world's 180,295 miles of generalized coastline—that works out to 6.5 percent, which makes the Coast Guard's responsibility look a bit less daunting.

TWO

DEFINITIONS AND DELINEATIONS

Several years ago an offhand remark won me a God's eye view of shoreline dynamics. When making hotel arrangements for a late-winter conference on geospatial technology for coastal managers, my name and phone number told the reservations clerk I was a member of the "honored guest" program—a fluke insofar as I hadn't stayed at the hotel since filling out its "free enrollment" form two years earlier. Asked whether I had any special requests, I blurted (not wanting to disappoint) that a room far from the elevator would be nice. The result was four days and three nights in a corner room with two balconies on the top floor of a nineteen-story hotel next to the ocean at Myrtle Beach, South Carolina.

I had seen the beach from hotel rooms before, but never like this. My first thought was regret for not bringing a camera. My second was mild relief in not having to obsess over the ideal field of view or the right moment to snap the shutter. I would have quickly filled an entire memory card before realizing the need for two additional photo sessions, about six hours apart, to capture the extreme positions of low and high tides.

If I had mapmaking in mind, a single photo shoot wouldn't do. As data for a nearby tide station confirmed, three days later the low-water line was lower, the high-water line was higher, and the intertidal zone was noticeably wider. Despite my short stay, Myrtle Beach provided a vivid example of how a gently sloping beach converts a

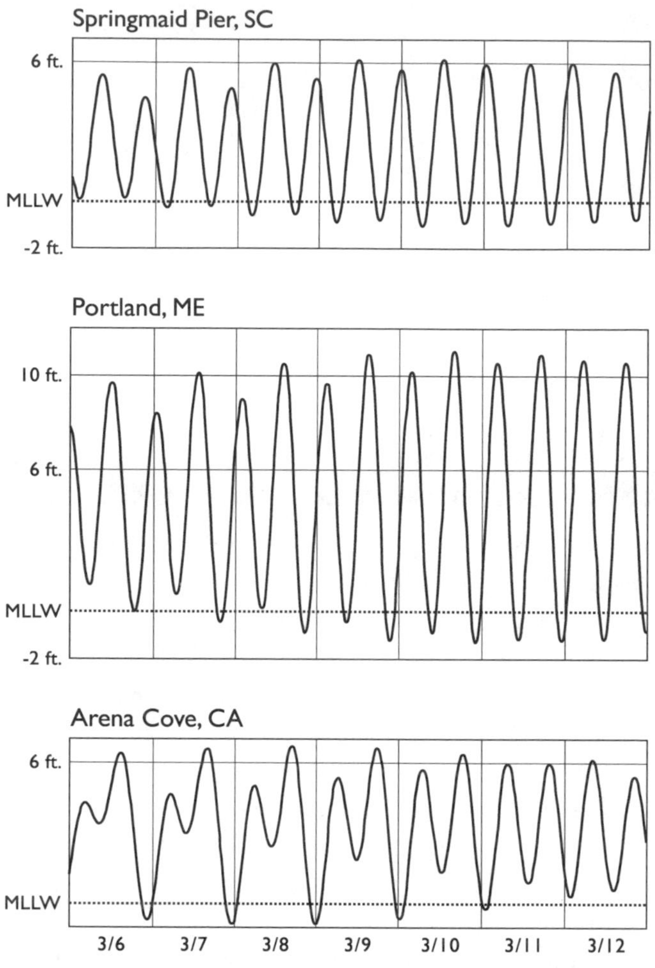

Figure 2.1. Predicted water level, in feet, above lower-low water, for selected tide stations, March 6–12, 2005. Redrawn from NOAA tide-prediction graphs based on Greenwich Time.

six-foot tidal range into a band much wider than the theoretical forty-foot corridor defined by a $\frac{1}{50}$-inch-thick shoreline on a 1:24,000 topographic map. It was also a useful reminder of coastal mapping's close ties to physics, astronomy, and measurement technology.

Coastal cartographers are haunted by complicated tides like those in figure 2.1, which describes a full week of sea-level fluctuations for two stations on the Atlantic coast and one on the Pacific.[1] Based on mathematical predictions calculated at least a year in advance from carefully calibrated measurements, these time-series graphs describe the pull of moon and sun on oceanic waters but ignore daily weather and tsunamis, both of which defy long-range forecasting.

The upper graph, for Springmaid Pier, South Carolina, near Myrtle Beach, shows not only the twice-daily rise and fall in water level but also the arrival of a new moon around 4 a.m., Greenwich Time, on March 10, when the more or less direct alignment of the moon and the sun with the earth maximized their gravitational impact. Because the moon is much closer to earth than our far more massive sun, its pull is greater by a factor of 2.5.[2] Both bodies pulling in the same direction produce a spring tide, characterized by lower lows and higher highs. Derived from the Anglo-Saxon word *springan*, meaning "to swell or bulge," the term has nothing to do with the seasons—spring tides arise twice a month throughout the year.[3] The "bulge" is apparent in the range between high tide and low tides, which increased as March 10 approached and declined thereafter.

High tides occur twice a day at Springmaid Pier, once when the moon is near its highest point above the horizon and again a half-day later, when the moon is on the opposite side of the rotating earth. This second tidal rise reflects the centrifugal force caused by our planet and its satellite spinning about a common center of mass. Centrifugal force is the outward pull one feels near the edge of a spinning carousel. Directed away from this common center of mass, 1,068 miles below ground on the moon's side of the earth's center, the centrifugal effect on tides is always greatest on the side of the planet directly opposite the moon. And because the moon is slowly orbiting the rotating earth, coastal areas typically experience two complete rising and falling tides every twenty-four hours and fifty minutes.[4]

I say *typically* because tides respond to diverse influences in different ways, as a close look at figure 2.1 will show. As all three tidal curves demonstrate, successive high tides are not always equal. At Springmaid Pier, for instance, the first high tide on March 6 was more than a foot higher than the next high tide, later that day. This first peak was higher partly because the moon was on the same side of the earth as the tide station and partly because the moon was also north of the equator. Whether the moon is above or below the

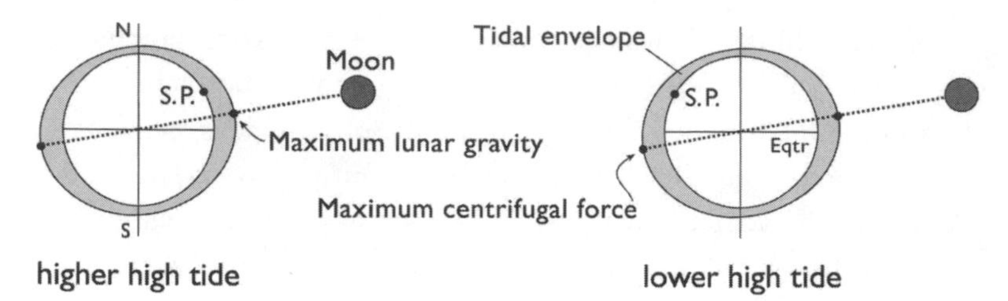

Figure 2.2. An exaggerated tidal envelope illustrates how a moon above the plane of the equator yields noticeably different water levels for successive high tides at Springmaid Pier, South Carolina.

equator on a given day is significant because lunar gravitation is greatest at the point directly below the moon.

The physics is easier to understand if you envision a worldwide tidal envelope with an elliptical profile pulled outward toward the moon by gravity and away from the moon on the opposite side of the earth by centrifugal force (fig. 2.2). When the sublunar point is in the northern hemisphere, as it was on the day in question, the moon's pull at Springmaid Pier is greater than if the sublunar point were in the southern hemisphere, thousands of miles away. But a half day later, when the moon is on the opposite side of the planet but still well above the equator (fig. 2.2, *right*), the point with the greatest centrifugal force is in the southern hemisphere and high tide at Springmaid Pier is noticeably below its previous level. This disparity diminishes as the approaching alignment of moon and sun dampens lunar dominance, as it did March 10.

Tidal predictions for Portland, Maine, reveal a more pronounced difference between successive highs earlier in the week. Although Portland's greater tidal range might be attributed (incorrectly) to its higher latitude, tides are influenced by the shape of the coastline, the configuration of the ocean floor, and the size of ocean basin. Because tidal forces are pulling not only upward toward the moon but also laterally, toward the sublunar point, the global tidal envelope endures massive horizontal shifts, particularly prominent in the Bay of Fundy, a long, funnel-like indentation in Canada's Atlantic coastline where the tidal rise can exceed fifty feet.[5] As figure 2.3 illustrates, a substantial drop in water level at low tide leaves boats resting on the bay floor, well below the wharf deck.[6]

Figure 2.3. Boats next to the wharf at St. Martins, New Brunswick, on the Bay of Fundy, are kept from tipping sideways at low tide by a wooden structure called a cradle, which rests on the bay floor to support the hull. Courtesy of Tourism and Parks New Brunswick, Canada.

Smaller, more confined water bodies like the Great Lakes and the Mediterranean Sea have markedly smaller tides, typically less than a foot, because moon and sun have much less water on which to work. By contrast, the comparatively vast Atlantic Ocean can register the full impact of lunar gravity and its centrifugal counterpart, while the Pacific Ocean is much too large for neatly uniform twice-daily lunar tides. Mixed tides like those described in the bottom graph for Arena Cove, California, a hundred miles north of San Francisco, are common in the Pacific Ocean, where successive low tides can be radically different.

Disparities like those at Arena Cove led the Coast and Geodetic Survey to base its charted soundings and bathymetric contours for the Pacific Coast on mean lower-low water (MLLW in fig. 2.1), computed as the long-term average of the lowest of each pair of low tides, instead of mean low water (the average of all low waters), in use along the Atlantic Coast.[7] As a safety measure, the mean lower-low-water datum—*datum* means "reference level"—helps mariners approaching a Pacific port make more reliable off-the-cuff estimates of whether, for instance, a ship with a twelve-foot draft is likely to clear an area with charted waters as shallow as twelve feet.

Mariners know to check a table of tide predictions whenever tight clearances are likely because of a new or full moon. As the graphs in figure 2.1 show, when moon and sun reinforce each other's gravity, water level at low tide can drop below the chart datum. Although a mean lower-low-water datum ensures that the chart indicates minimum depth most of the time, tide tables giving the time and water level for every high and low tide are especially valuable when a navigator must await a rising tide in order to clear a reef or sandbar blocking a harbor.

Why not, just to be safe, reference charted depths to the lowest level the water is ever likely to drop because of gravity and earth's rotation? That's the strategy recommended by the International Hydrographic Organization (IHO) and adopted by Britain's Admiralty Hydrographic Office and the U.S. military.[8] A bit too low, though, for American civilian chart makers, who thought the lowest astronomical tide would make perfectly navigable channels look overly shallow. Charts, they reasoned, should reflect comparatively normal conditions. Make navigation too easy, and mariners might start believing the water could never drop below the lowest astronomical tide—a dangerous view because unusual weather can produce low tides below the predicted astronomical minimum.

In 1980 the NOS compromised, sort of, by adopting mean lower-low water as the standard chart datum for all the nation's coasts.[9] In most cases the shift made little difference. At Portland, Maine, for instance, mean lower-low water is only 0.35 feet below mean low water, while at Springfield Pier, South Carolina, the difference is only 0.19 feet.[10] Even so, astronomical minima respectively 3.45 and 3.89 feet below the chart datum underscore the need for reliable tide tables.

Because tide tables are an indispensable adjunct to large-scale nautical charts, hydrographic agencies have assumed responsibility for tide prediction, which requires precise measurement of water level throughout the day over many years as well as a theoretically sound, computationally practicable method for projecting tidal behavior a year or two into the future. Although astronomical tables that predated movable type allowed for crude tide tables based on phases of the moon, precise records of water level were largely lacking until the mid-nineteenth century, when the clock-driven tide gauge replaced the tide staff, essentially a vertical ruler strapped to a pier.[11]

Isaac Newton laid out the physics of tidal forces in his *Principia Mathematica,* published in 1686. Two centuries later William Thompson translated Newton's theory into a computational breakthrough of nearly equal significance to hydrographers. In 1868, at a meeting of the British Association for the Advancement of Science, Thompson (better known today as Lord Kelvin) proposed reducing "the complicated motions of the tides into a series of simple harmonic motions or waves in different periods and with different amplitudes, or ranges [that when] added together [would] give the aggregate tide."[12] Extending a notion pioneered by French mathematician Pierre-Simon Laplace, he represented the tidal influences of sun and moon by ten theoretical satellites, all in the same plane, all moving at constant (but different) speeds in circular orbits around the earth, and all amenable to modeling by a precisely fabricated mechanical computer.[13] "To substitute brass for brain in the great mechanical labour of calculating the elementary constituents of the rise and fall of the tides,"[14] he designed a Tide Predictor to simulate the combined effect of tidal influences. Built in 1873 with British Association funds, Thompson's analog computer proved the concept of mechanical tide prediction and led to similar devices, with additional hypothetical satellites to account for less regular tides in India and North America.

William Ferrel, better known for his contributions to meteorology, designed the first American adaptation, which modeled nineteen tidal components.[15] Completed in 1882, Tide Predicting Machine No. 1 was the source of tidal predictions published from 1883 to 1912, when Tide Predicting Machine No. 2 took over. The 2,500-pound apparatus was eleven feet long, two feet wide, and six feet tall (fig. 2.4).[16] As an operator turned a crank, its gears, chains, and pulleys summed the tide's harmonic influences and plotted predicted wa-

Figure 2.4. Front view of U.S. Coast and Geodetic Survey Tide-Predicting Machine No. 2. The crank is to the left of the operator's desk, above which dials indicating the day (outer), hour, and minute resemble a giant face. From the NOAA Center for Operational Oceanic Products and Services Web site, http://tidesandcurrents.noaa.gov/images/mach1b.gif.

ter level on a continuous roll of paper. With predictions based on as many as thirty-seven tidal harmonics, the unit was more flexible and accurate than its predecessor. According to a 1916 gloss on coastal charting, the machine, "when set up for a particular station and some future year, takes into account [not only] the tide-producing forces of the sun and moon [but also] the modifications of their effects by the conditions peculiar to the particular station."[17] The set-up for each station could take an hour or two, but in a day or less an experienced operator could tabulate "on a form ready to go to the printer, the day, hour, and minute, and the feet and tenths of each high and low water in one year."[18] During World War II, TPM-2 operated around the clock, providing tidal forecasts for Allied invasions.[19] Retired in 1965, when a computer took over, it occupies a glass enclosure in an obscure science exhibit at NOAA headquarters in Silver Spring, Maryland.[20]

Besides helping navigators avoid dangerously shallow water at low tide, predicted water levels are useful in planning aerial surveys for coastal cartography.[21] For example, a flight timed for the narrow window from shortly before to shortly after the forecast arrival of mean lower-low water can help chart makers freeze on film the position of the chart's vertical datum, to which soundings and bathymetric contours refer. As an aid to mariners, nautical charts show this line as the seaward edge of the intertidal zone, also called the foreshore.[22] The State Department is particularly interested because lower-low water defines the baseline for the territorial sea, which extends twelve nautical miles outward to a crucial international boundary.

Carefully timed aerial photography can also capture the mean high water line, sometimes approximated by the seaward limit of vegetation. As the only official shoreline on nautical charts, the line of mean high water gives mariners a visually recognizable boundary between land and sea. And it's often found on topographic maps because the U.S. Geological Survey usually takes its shorelines from National Ocean Service charts.[23] Tide-coordinated aerial images also help mapmakers define the mean higher-high-water line, yet another shoreline, which is used to separate private and state land in Hawaii, Louisiana, and Texas.[24]

The first two delineations are legally significant in most coastal states. Particularly important to financially strapped governors and legislatures, mean lower-low water is the baseline for the three-mile seaward boundary within which individual states control mineral leases and seabed mining.[25] By contrast, mean high water defines not only the national cartographic shoreline but also the legal boundary between private and public land in most coastal states. Exceptions include Delaware, Massachusetts, Maine, Pennsylvania, and Virginia, where private ownership can extend as far seaward as the line of mean lower-low water.[26] As chapter 9 observes, environmental regulations, flood insurance, and setback restrictions on new construction call for diverse definitions of the shoreline.

Topographic maps further complicate the picture by referencing their elevation contours to mean sea level, a blatant contradiction of the widely held but erroneous belief that the cartographic shoreline, defined by mean high water, is the zero-elevation contour. It's also dangerous to assume that the mean sea level used as a fixed vertical reference for most topographic maps published since the 1930s is the same as today's mean sea level, which reflects the cumulative effects of rising seas and coastal subsidence, occasionally hastened by withdrawal of underground oil and natural gas. This naiveté proved particularly devastating for New Orleans in 2005, when levees along the Inner Harbor Navigation Canal failed during Hurricane Katrina. As an Army Corps of Engineers post mortem discovered, the levees were two feet lower than they should have been because flood-control engineers confused the National Geodetic Vertical Datum of 1929 with mean sea level.[27]

Equally ornery are nautical charts, which relate elevations of lighthouse beacons and other landmarks to mean high water but sometimes reference their elevation contours to mean sea level.[28] Subtle differences between American charts and those of other nations arise as well, for example, where bridges or power lines span navigable waterways. The IHO lets chart makers relate vertical clearances to mean sea level, but the United States uses the more conservative datum of mean higher-high water—an extra measure of safety for skippers of tall ships who fail to check tide predictions and weather forecasts.[29]

Tide prediction and the various legal and cartographic shorelines share a common database, collected over a nineteen-year period hydrographers call a tidal epoch. The tide predictions in figure 2.1, for instance, are based on data collected from 1983 through 2001, the same epoch used to calculate water-level averages for the three stations shown. Nineteen years of measurement is necessary to cover the full range of relative distances among earth, moon, and sun.[30]

Where reliable data are available, estimates of sea-level change are based on much longer periods. At Portland, Maine, for instance, data for the period 1912 to 1999 indicate a long-term rise in mean sea level of 1.91 mm per year, or 0.6 feet per century, less portentous than the rise of 1.7 feet per century projected for Springfield

Pier.[31] Although the latter rate is based on observations initiated in 1957, similar trends at nearby stations with a longer record suggest a plausible threat to South Carolina's gently sloping coastal plain. I could not find a comparable analysis for Arena Cove, California, but a 0.82-feet-per-century rise at Point Reyes, seventy miles south, and a 0.16-feet-per-century decline at Crescent City, two hundred miles north, attest to the comparative complexity of shorelines on our tectonically active Pacific coast.

Long-term water-level records are not the only evidence for rising seas resulting from climate change or vertical movement of the earth's crust. Then-and-now comparisons based on old maps can confirm ominous trends, especially when detailed historic charts are based on systematic surveys of the high-water line, our national shoreline datum since the federal government began mapping coastal waters in the mid nineteenth century. Congress authorized the Survey of the Coast in 1807, but political wrangling, the War of 1812, and puny budgets delayed a serious start until the 1830s.[32]

Because early coastal charts show only what their surveyors could infer from dunes, vegetation, and other shoreline indicators, our historic coastline is an "interpreted shoreline," subject to seasonal perturbations and the transient effects of severe storms.[33] But so too are modern delineations supported by exact calculations of mean high water at tidal stations. The challenge today is no less daunting than in 1928, when the Coast and Geodetic Survey's *Topographic Manual* warned, "Where the beach slopes gently and the tidal range is large, it is practically impossible to identify exactly the mean high-water line."[34]

Impressed with the professionalism of Coast Survey cartographers and their attention to detail, the National Ocean Service scanned its predecessor agency's maps known as T-sheets—*T* for topographic—and traced their shorelines into an electronic database for use by coastal scientists concerned with erosion and sea-level rise.[35] The database also includes historic aerial photography, used in the compilation of T-sheets from the 1940s onward. Drafted as manuscript maps for use in compiling finished nautical charts, the T-sheets describe coastal roads, prominent structures, and visible rocks as well as the shoreline at a scale larger than that of the published chart.

Because of the shoreline's relevance to mariners, surveyors focused on getting it right. Apparently they succeeded. Stephen Leatherman, a well-known coastal scientist, calls "NOS 'T' sheets... the most accurate maps commonly available for the coastal zone."[36]

Encouraged by this endorsement, I wondered whether T-sheets could help solve a mystery I call the case of the missing fifth island. Recall that the USGS map showing Five Islands, Maine (fig. 1.1) names only four nearby islands. I suspected that the fifth island was near Crow Island, the most northern of the four, because the 1874 *Coast-Pilot,* a government guide to navigation channels and harbors, not only describes the islands as "all small, and five in number" but also notes that "the *two* northernmost lie close together, about four hundred yards to the northward of the [two] middle islands."[37] What's more, "they are both low and very small, and lie closer to the shore than any of the others." The 1879 edition offers a similar description, while previous and later versions dutifully mention the village and its harbor but ignore its offshore namesakes. Even so, the USGS map and a nautical chart for the same area (fig. 1.2) show a Crow Island with a pair of markedly smaller satellites, one on the bay side and the other near the mainland. Could one of them be Five Islands' fifth island?

NOAA's Coastal Service Center, in Charleston, South Carolina, answered my initial query with a pair of index maps, one outlining areas covered by T-sheets surveyed between 1850 and 1866 and the other describing coverage by photo-based surveys from the 1940s and early 1950s. Each index yielded one T-sheet that included Five Islands. Mike Rink, who oversees the data, retrieved the scanned images and placed copies in a "file transfer" folder, from which I downloaded the maps. Despite fuzzy labels and blurry symbols, these low-resolution scans reveal noteworthy change in island shorelines.

Surveyed in 1861, the older of the two historic maps (fig. 2.5, *left*) describes a tiny village near a cluster of islands. Double-dash lines delineate dirt roads, tiny black symbols with squared corners show building footprints, and thin dot-dash-dot-dash lines with squared corners represent fences or stonewalls—vintage coastal New England. The map reflects a natural environment much like it is to-

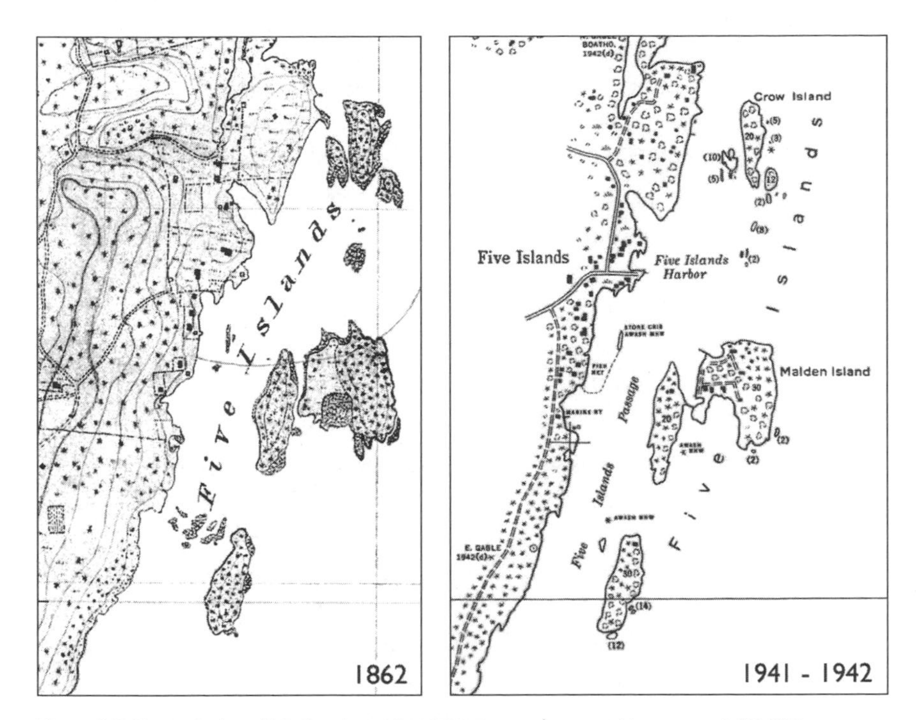

Figure 2.5. Excerpts from U.S. Coast and Geodetic Survey topographic manuscript T-889 *(left)*, drafted at 1:10,000 from an 1862 survey, and Coast and Geodetic Survey shoreline survey manuscript T-5988 *(right)*, compiled at 1:10,000 from October 1941 aerial photography and field checked in 1942. Courtesy NOAA Coastal Services Center.

day. Gently flowing elevation contours encircle a low hill just west of the hamlet, clusters of short horizontal dashes with protruding reeds represent marshland, and tiny, more or less evenly spaced asterisks indicate pine forest. Because a different field party would have recorded soundings and other offshore details on a separate manuscript, the detailed shoreline stands out as a sharp boundary between a filled-in landscape and a featureless seascape.

If nameworthiness depends on size, the older T-sheet suggests the inner, southwestern satellite of Crow Island is the elusive fifth island. Clearly the fifth biggest island in the immediate area, it's also closer to the shore and thus more noticeable than the small islet immediately southeast of Crow Island. According to Lynne Jones, president of the local historical society, long-time residents call it Little Crow Island.[38] Although maps suggest the area might more accurately be labeled Four Islands or Six Islands, early settlers named it after the number of islands readily visible from the dock.

The 1941–42 T-sheet (fig. 2.5, *right*), based on an aerial survey and a subsequent field check, shows a much-diminished Little Crow Island, with a peak elevation ten feet lower than the highest point on Crow Island and two feet below the high point on its more seaward competitor. Careful comparison of the two maps reveals other changes in the local shoreline, at least partly the result of eight decades of wave action and rising seas. Also apparent are the hamlet near Five Islands Harbor and the appearance of summer homes on Malden Island, where elevations reach thirty feet. Aerial surveying was especially adept at estimating island elevations, of interest to mariners seeking prominent landmarks and homebuyers wary of coastal flooding.

While Little Crow Island is noticeably larger on more recent maps, it's still a risky place to build a home. At least that's an impression readily drawn from the flood hazard map (fig. 2.6) issued after Georgetown, Maine, which includes Five Islands, joined the National Flood Insurance Program. Gray shading marks the "Special Flood Hazard Area," inundated by the hypothetical "100-year flood," with a one-percent chance of occurring in any given year. Although the term implies a once-a-century event, flooding this extensive can strike two years in a row, or even twice in one year. In this sense, the gray shading defines a shoreline especially important to home owners because banks will not mortgage property in flood zones unless the owner buys flood insurance. As the map excerpt shows, most of Little Crow Island lies in Zone AE (EL 10), defined by a flood elevation of ten feet above the "National Geodetic Vertical Datum"—another shoreline, established in 1929 and not to be confused with today's mean sea level.

Elevation is critical. Flood insurance is more expensive in Zone AE than in Zone X, above the 100-year-flood line, but landowners can lower their insurance rates by building houses on stilts above the "base flood level," a common practice in the Carolinas, where the flood zone is much broader than in Maine. Potential damage and insurance rates are greater still in Zone VE (EL 16), on the east side of Crow Island and all around its little neighbor to the east.

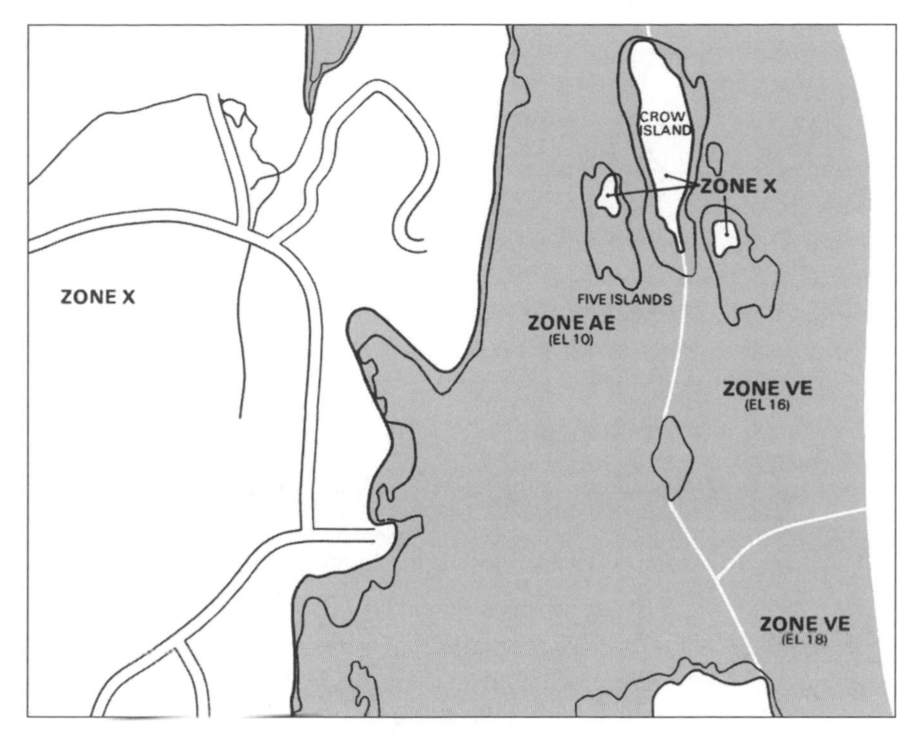

Figure 2.6. Excerpt from the Federal Emergency Management Agency's Flood Insurance Rate Map for Georgetown, Maine, community-panel number 230209 0005 B, May 17, 1988.

V stands for "velocity hazard," a warning that the unwary buyer risks losing the lot as well as the house to brutal waves. To compensate, the Federal Emergency Management Agency (FEMA), which assesses risk, set a higher base elevation (sixteen feet, rather than ten) for the east side of Crow Island. Because the elevation data used to fix the flood-hazard shoreline are suspect, FEMA recently undertook the massive task of replacing its old flood maps, based on inadequate elevation data derived from 1:24,000 USGS topographic maps.[39] Some alteration of Little Crow Island's flood-hazard shoreline seems inevitable.

The following chapter emphasizes the difficulty of plotting the first coastline from offshore and the complications arising from a vague at best sense of one's longitude.

THREE

NEW WORLDS AND FICTITIOUS ISLANDS

In 2003, after a decade of delicate negotiations and intense fund-raising, the U.S. Library of Congress paid a German prince $10 million for a 1507 world map proudly exhibited beneath the banner "America's Birth Certificate."[1] Half the money came from private donors and half from the taxpayers. Gaining congressional approval was less difficult than arranging the export of a rare artifact on Germany's National Register of Protected Cultural Property.[2] The sole survivor of perhaps a thousand printed copies, the map surfaced in 1900 at Wolfegg Castle, in Bavaria.[3] Compiled by Martin Waldseemüller, a scholar working in Saint-Dié, in northeastern France, on an updated edition of Ptolemy's *Geography,* the map reflects published accounts and secret diaries of Christopher Columbus, Amerigo Vespucci, and other explorers. Columbus, who died in 1506, insisted he had reached Asia, but Vespucci saw the New World as a separate land mass. Waldseemüller agreed and boldly inserted "America" toward the bottom of a skinny South American continent, clearly not part of Asia, in the map's lower-left corner (fig. 3.1). In the 1530s, the name took hold for North America as well.

Emblematic of Renaissance efforts to chart the world's coastlines, Waldseemüller's map is impressive, even when it's just a full-size facsimile like that on display in 2005, when I visited the Jefferson Building, across the street from the Capitol. Printed in twelve sections from printing plates engraved in wood, the map is four and a

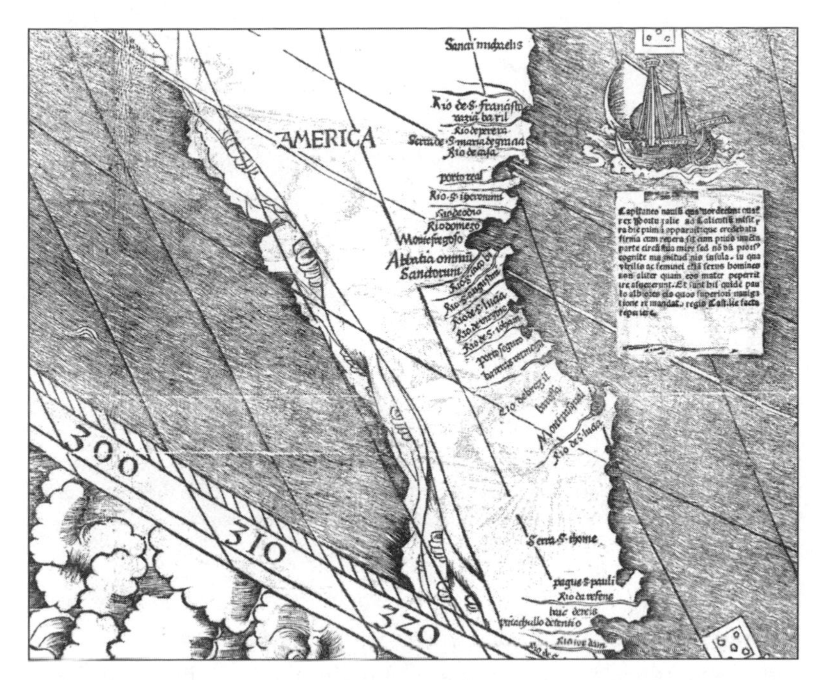

Figure 3.1. Portion of lower-left panel of Martin Waldseemüller's 1507 world map with the name "America" in the approximate location of present-day Paraguay. From the Library of Congress American Memory Web site, http://memory.loc.gov/ammem/.

half feet tall and eight feet wide, a little bigger than a panel of sheet rock from Home Depot. Eager to share their new acquisition more widely, library officials placed a high-resolution image of the original on their American Memories Web site, where it could be downloaded as a twenty-seven-megabyte file by anyone with a high-speed Internet connection or a great deal of patience. As the excerpt in figure 3.1 suggests, the Web version reveals a grid of meridians and parallels, a sprinkling of ornaments and text blocks, and the striking contrast of an Atlantic coast penetrated by named rivers and a Pacific coast rendered in vague flourishes.

To appreciate the map's importance, viewers must step back about twenty feet and examine its distinct western hemisphere, with northern and southern American continents separated from Asia by a vast Pacific Ocean—a revolutionary worldview I've tried to summarize in much reduced form in figure 3.2. Waldseemüller's take on the continents proved a lucky hunch. In 1507, thirteen years before

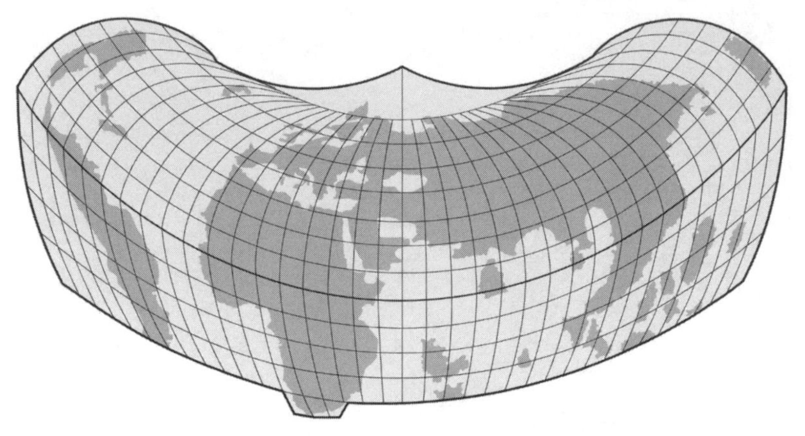

Figure 3.2. Generalization of the Waldseemüller map's coastline and spherical grid. Only larger islands are included. Traced from an image downloaded from the Library of Congress American Memory Web site, http://memory.loc.gov/ammem/.

Ferdinand Magellan sailed into the Pacific through the strait now bearing his name, the notion of an enormous ocean between Asia and the Americas was highly speculative but no less credible than Columbus's belief that the islands of the West Indies were on the fringes of Asia—a theory that started to crumble when his descriptions of Cuba and Española (the island now shared by Haiti and the Dominican Republic) failed to match Marco Polo's accounts of the Spice Islands, which Columbus had hoped to reach by sailing westward.

While Waldseemüller's vision of separate American continents ultimately proved correct, it was no more plausible in the early 1500s than a competing notion of the Americas as a large subcontinent extending southeastward from northern Asia, to enlarge the Indian Ocean at the expense of the Pacific. Oronce Fine, a French cartographer, reinforced this view by stretching a much smaller Asian peninsula on Ptolemy's map eastward into the Americas, thereby turning southern China into Mexico.[4] Eager for acceptance, Fine, Waldseemüller, and their contemporaries mixed decoration and delineation. As cartographic historian Thomas Suárez points out in *Early Mapping of the Pacific,* "mapmakers were forced to step back to a more medieval approach, mapping from inference and camouflaging the uncertainty with vignettes, yet at the same time marking latitudes as though geographic placement had some meaning."[5]

Waldseemüller's 1507 opus demonstrates the persuasive power of maps that are widely distributed and prominently displayed. Even if the thousand-copy estimate of its circulation is only half correct, as the first printed map of the New World it was the cartographic best-seller of its day. Formatted for public display, it gained the attention of other mapmakers, who mimicked its coastlines and place names. Visibility had a downside, though, insofar as paper deteriorates with prolonged exposure to light. The copy at Wolfegg Castle was fortunate: bound in a book with other map sheets rather than mounted as a wall map, it escaped certain deterioration.

So convincing were the 1507 map's lines and labels that Waldseemüller was unable to stem its growing acceptance when he retracted his view of the New World by publishing a substantially revised world wall map, the *Carta Marina*, in 1516.[6] More in accord with Columbus than Vespucci, who died in 1512, his new map presents North America as an extension of Asia and replaces the seminal "America" with "Terra Incognita" (unknown land). The formerly enormous Pacific Ocean now ends abruptly at the right edge of the map and does not appear at all on the left, while upper South America becomes "Terra Nova" (New World), and North America emerges as "Terra de Cuba," even though the island now bearing the name remains separate. Advertised by the Library of Congress as "the first printed navigational chart of the entire world," the *Carta Marina* fetched only $4 million in 2004, when a private donor bought it for the library from the same German prince.[7] Who says names don't matter?

California provides another example of the persuasive power of fictitious coastlines based on speculation or hearsay. In 1622 English mathematician Henry Briggs, a prominent advocate of logarithms, dusted off a theory attributed to early-sixteenth-century conquistador Hernán Cortés, whose explorations of Mexico and Central America led him to conclude that California, or at least Baja California, was an island.[8] Although Mercator and other influential sixteenth-century mapmakers had largely put Cortés's assumption to rest, it appealed to Briggs, who believed strongly in the mythical Northwest Passage and eagerly endorsed the existence of similarly convenient passages elsewhere. In an appendix to a friend's book on the British colony in Virginia, the respected mathematician cited a newly disclosed map "brought out of Holland" as evidence that California "is now found to be an island stretching it selfe from 22 degrees to 42, and lying almost directly North & South."[9]

Hardly a secret, the map in question had appeared on the title page of a travelogue published in Amsterdam earlier that year. It was apparently based on a sketch map drawn in America by a Spanish friar in 1620 and intercepted by the Dutch.[10] Its author, Father Antonio Ascensión, seems to have misread reports by Juan de la Fuca and Martin d'Aguilar, who explored the California coast in 1592 and 1602, respectively. A willfully egregious misreading in the opinion of Dora Beale Polk, who in *The Island of California: A History of the Myth* labeled him "an impressionable simpleton turned crackpot."[11] Among the good padre's excesses is a memoir in which he calls the mythical passage "the Mediterranean Sea of California, since it is between lands so large and extended, [and] must be about fifty leagues wide."[12]

Inspired by Briggs's endorsement, mapmakers eagerly detached California from the mainland.[13] Abraham Goos, a Dutch engraver, grabbed the first seat on the bandwagon with a 1624 world map that lopped off the northwest corner of North America. English mapmaker William Grent climbed aboard with a 1625 world map showing a jalapeno-shaped California, separated from the mainland by a passage five degrees wide and tapered southward from a flat northern coast 10 degrees wide at 42°N to a rounded tip at 22°N.[14] In addition to affirming Briggs's description of the island, Grent's map supported the Northwest Passage myth with a fictitious chain of bays extending from ocean to ocean through what's now northern Canada. Unable to observe the California coast directly, other mapmakers deferred to Briggs and his followers. Map historian Ronald Tooley found an insular California on an even hundred maps published between 1625 and 1770.[15] Further searching would surely have discovered others.

Not all the detached Californias inhabit world maps. The Library of Congress owns a regional sailing chart produced by Dutch mapmaker Joan Vinckeboons around 1650 (fig. 3.3) that is not on Tooley's list. Enhanced by a compass rose and a network of straight-line sailing directions called rhumb lines, the chart invites mariners to approach California from any direction or navigate a

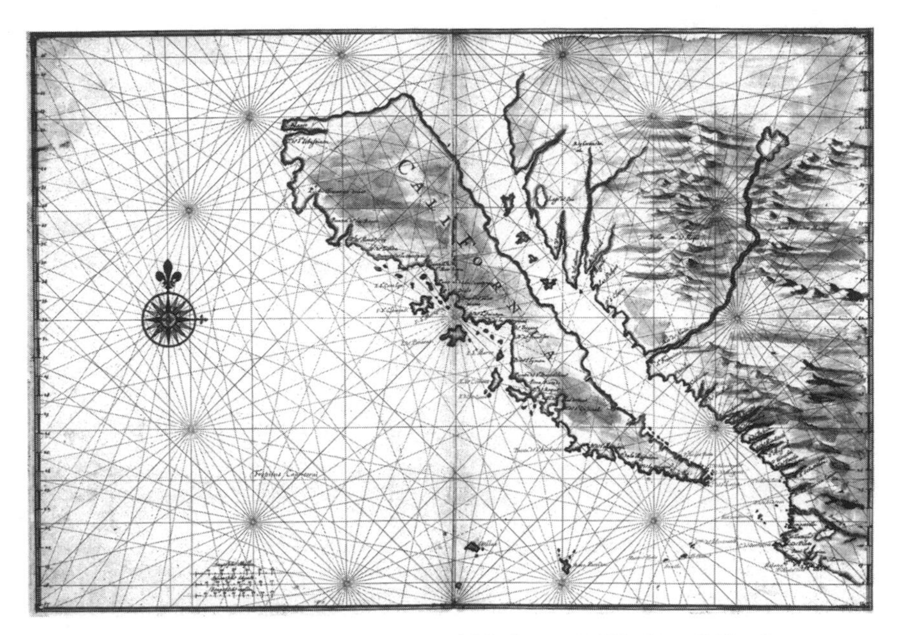

Figure 3.3. Sailing chart by Joan Vinckeboons, published around 1650, shows California as an island. From Library of Congress American Memory Web site, http://memory.loc.gov/ammem/.

broad channel on their way from Mexico to northern China. With little justification, Vinckeboons appropriated a cartographic framework pioneered in the fourteenth century by portolan sailing charts, so called because chart makers based them on books of sailing directions called *portolani*. Unlike Vinckeboons's mythical coastline, the portolan charts of the Mediterranean were comparatively accurate in their delineations, and characterized by closely spaced place names, extending inland from a perpendicular coastline.[16]

Ascensión's hoax—what else would you call it?—persisted well into the eighteenth century. Tooley identified French geographer Guillaume Delisle as a leader of the resistance.[17] Averse to unsubstantiated theories, Delisle merely left blank those parts of the map lacking reliable verification. Eusebio Kino, a Jesuit priest skeptical of Ascensión's claims, confirmed the account of a Spanish official who had walked from Santa Fe, New Mexico, to the Pacific.[18] Kino's map, printed in 1705, contradicted prevailing wisdom and was not well received.[19] Between 1715 and 1730 Dutch map publisher Pieter Van

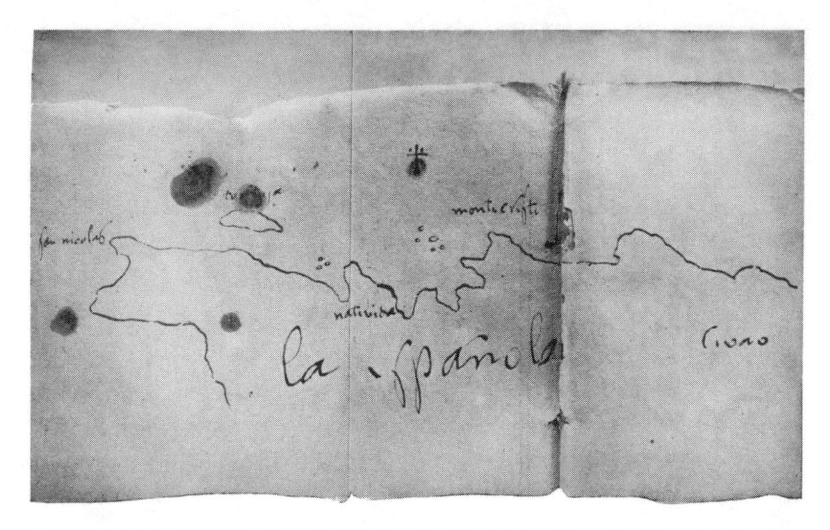

Figure 3.4. Northern coast of Haiti as sketched by Christopher Columbus in 1493, on the first of four voyages to the New World. Facsimile from John Boyd Thatcher, *Christopher Columbus*, 3:89.

der Aa, who apparently waffled, proved himself a shrewd businessman by offering two maps, one showing California as an island and the other presenting it as a peninsula.[20] True believers held fast until 1746, when another Jesuit, Fernando Consag, rowed up the Gulf of California and searched in vain for a northern outlet to the Pacific.[21] The following year the Spanish sovereign Ferdinand VII settled the matter by decreeing, "California is not an island."[22]

When not promoting a theory or borrowing another cartographer's delineations, mapmakers compiled coastlines of little-known areas from explorers' sketches like figure 3.4, a rendering of the northeastern coast of Española attributed to Christopher Columbus. Considered a rare specimen of the Great Admiral's handwriting because of its six place names, the map surfaced around 1894 among a collection of manuscripts that the Duchess of Alba purchased through a third party from the widow of a librarian who might have stolen them from his employer.[23] In 1902 the duchess published an annotated collections of manuscripts in Columbus's handwriting that included the map.[24]

Despite doubts about its authenticity,[25] facsimiles of the drawing decorate biographies of Columbus as well as several cartographic histories. Some reproductions are sharp, cleaned-up versions, while others, like figure 3.4, show the outline of facing pages in the admiral's notebook and the deterioration of its paper. This example is from a 1904 biography by John Boyd Thatcher, who called it "the first map made of any portion of the New World."[26] Thatcher considered the drawing not only authentic but a significant find. So did eminent historian Samuel Eliot Morison, whose close reading of Columbus's own account of the first of four voyages to the New World indicates the sketch was made in early 1493, not long after the ill-fated Christmas Eve 1492, when the expedition's flagship *Santa Maria* hit a reef near the middle of the map, little more than a mile offshore.[27]

Morison was not a typical academic historian. In addition to four decades on the Harvard faculty, he served in the U.S. Navy during World War II and retired with the rank of Rear Admiral. A gifted writer and naval historian, he believed in supplementing archival research with visits to historically significant sites.[28] For fuller insight to Columbus's skill as an observer and navigator, he relied on the explorer's journals and vessels of similar size to retrace the historic voyages across the Atlantic and around the Caribbean.

Particularly important to map scholars is Morison's January 1939 reconnaissance of Haiti's northern coast, which validated Columbus's rough sketch map and yielded a detailed cartographic summary of the explorer's movements 446 years earlier. An old money Bostonian whose connections included the Episcopal bishop of Haiti, Morison persuaded the Haitian army to provide a coastguard cutter and crew. Using this "seaworthy little motor cruiser" on the water and a car and mules on land, he reconstructed Columbus's eastward journey along the coast and plotted significant sites, including the reef where the *Santa Maria* suffered "one of the most notable shipwrecks in history" and the likely location of Navidad, "the first European settlement in the New World," where the exhausted expedition paused ten days to regroup before sailing back to Spain.[29] Loss of the *Santa Maria* forced Columbus to build a fort there and leave behind forty men—he returned eleven months later to find the fort burned and all the men dead.

Sketching a coastline from memory, even one explored just a few weeks earlier, hardly guarantees geometric accuracy, especially when it's difficult to maintain uniform scale. As Morison noted, Columbus had a functional compass, which he used skillfully, but no reliable way of measuring distance at sea.[30] For rough approximations, he resorted to dead reckoning, an ominous term for guessing the ship's speed and multiplying by the elapsed time, estimated no less crudely with a half-hour glass, which had to be watched carefully and flipped at the right moment. Some navigators would toss a log overboard, and estimate the speed at which the ship moved away, but not Columbus, who usually overestimated distance.

Making allowances for technological shortcomings, Morison found the explorer's cartography remarkably reliable, at least in context, and superior to many later efforts. Tortuga, for example, was "far more accurately drawn on Columbus's own map than on any other before the eighteenth century."[31] Named for its apparent resemblance to a turtle, Tortuga is the narrow, linear island just off the coast in figure 3.4. Its comparatively correct portrayal surely reflects multiple encounters, reconstructed from the explorer's journal, as he zigzagged eastward through the channel between Tortuga and Española.

Less tolerant of twentieth-century mapmakers, Morison complained that "no very accurate maps of northern Haiti exist." Particularly impoverished was the area near Navidad, where the coastline "has never been properly sounded or surveyed and . . . is very inaccurately represented on the most modern United States Hydrographic Chart."[32] This latter assessment probably refers to the soundings, hazards, and other navigational details of the 1930s nautical chart that became his expedition's base map. Despite its shortcomings, the chart offers a reliable portrait of both the general coastline and the general tidal shoreline, as discussed in chapter one.[33] At least that's my impression after comparing Morison's excerpt from the offending chart with the coastline portrayed on relatively recent Central Intelligence Agency maps of Haiti and the Dominican Republic.

Curious about shapes and distances, Morison juxtaposed Columbus's drawing with the relevant portion of the 1930s nautical chart and a corresponding excerpt from a world map drawn around 1500 by Juan de la Cosa, official cartographer on the explorer's second voyage.[34] Although side-by-side comparison revealed that both

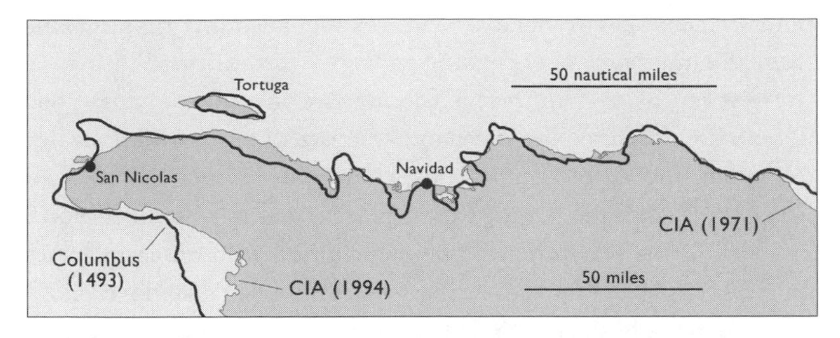

Figure 3.5. Columbus's sketch of the coastline superimposed on a shoreline traced from Central Intelligence Agency maps of Haiti and the Dominican Republic published at 1:1,000,000 in 1994 and 1971, respectively. The Columbus coastline, traced from figure 3.4, was rotated and scaled for a close visual fit. Place names are from the Columbus drawing.

early maps exaggerated the coast's capes and bays, Cosa's rendering was more extreme but understandably so. Because of difficulties in measuring distance, explorers' maps often magnify coastal features where the shoreline is intricate or the expedition lingers—an effect similar in appearance to the purposeful distortions on modern tourist and subway maps. And on a map compiled months or years later from its compiler's sketches or another observer's notes, prominent indentations or protrusions can grow larger still. What's more, a mapmaker relying on off-the-coast drawings by multiple observers can too easily resolve discrepancies with duplicate depictions of the same feature in different places.[35]

For a more focused evaluation, I superimposed a tracing of Columbus's sketch on the CIA's coastline, and scaled and jiggled the former to fit the latter. The result (fig. 3.5) confirms the exaggeration of features important to the expedition. On the western end of the coast, for example, Columbus greatly enlarged the harbor at San Nicolas (now Môle Saint Nicolas), where he anchored. Overstated capes and bays are readily apparent farther eastward, particularly around Navidad. Even so, the explorer seems to have captured the salient trend of the northern coastline, which he examined firsthand from San Nicolas eastward. The greatest discrepancy occurs in the southwestern part of the map, which shows an area glimpsed in the distance but not visited. It sounds trite, but being there and taking good notes almost always yields a more reliable coastline than looking from afar, guessing, or depending on vague, unconfirmed theory.

However helpful in reconstructing a coastline's shape, good notes were of limited use if you didn't know where you were, or like Columbus and his contemporaries, knew your latitude but not your longitude.[36] Latitude, defined as a place's angular distance north or south of the equator, could be determined with an astrolabe or quadrant, a device for measuring a celestial body's angle of elevation above the horizon. At night, the elevation of the polar star conveniently equaled latitude. During the day, a navigator could track the sun's ascent, measure its maximum elevation, and look up the latitude in an astronomical table for that particular day. Longitude might have been figured with equal ease if the ship had a precision timepiece called a chronometer showing the current time at the prime meridian, the line of zero longitude. Because the earth rotates through a full circle (360 degrees) in twenty-four hours and covers 15 degrees in a single hour, a time difference of four minutes represents a 1-degree difference in longitude. If it's noon locally—you know because the sun has reached its highest elevation, and is about to descend—and if your chronometer tells you it's 12:08 p.m. at the prime meridian, your longitude is 2°W. The calculation is straightforward but only as reliable as the chronometer. If it lost only twelve minutes since you set sail six months ago—a little less than four seconds a day—your longitude would be off by 3 degrees.

Columbus was unaware of time's value as a surrogate. Flemish astronomer Gemma Frisius didn't propose using portable clocks to estimate longitude until 1530,[37] and Swiss clockmaker and mathematician Jost Burgi didn't introduce the second hand until 1600.[38] Although the predicted times of lunar eclipses and other astronomical events published in an ephemeris (celestial almanac) would eventually be used to calculate longitude, sixteenth-century telescopes were too primitive for celestial navigation.[39] Forced to rely on latitude as their primary signpost, Columbus and later explorers resorted to parallel sailing—heading due north or due south to the parallel of latitude through their destination, and then sailing east or west to their objective.

By the end of the sixteenth century explorers and traders had

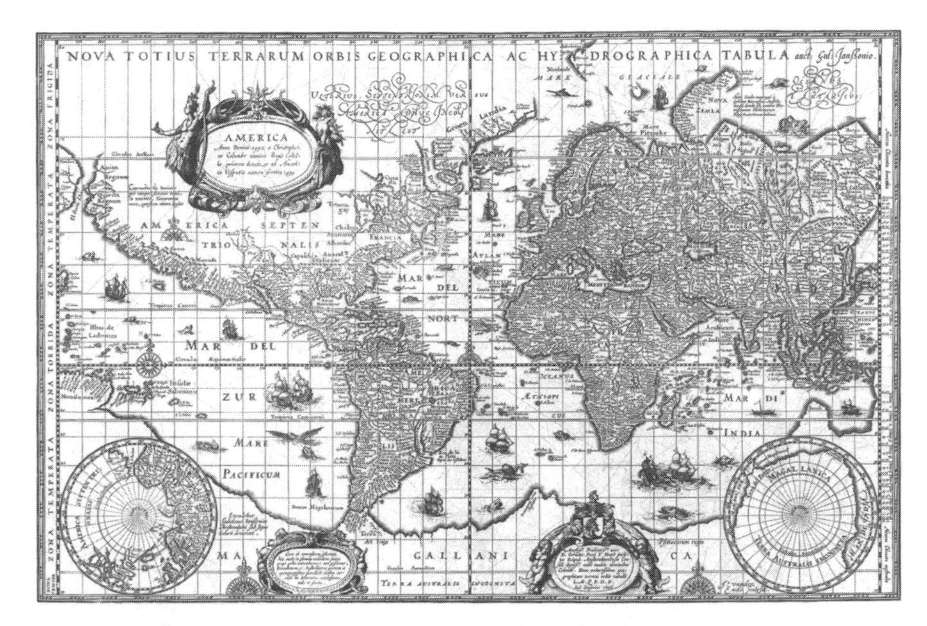

Figure 3.6. Willem Janszoon Blaeu's 1606 world map. Original map measures 12 by 18 inches (30 by 45 cm), including an elaborate border removed in this view. From the Library of Congress American Memory Web site, http://memory.loc.gov/ammem/.

what Columbus lacked: sailing charts showing how to reach the New World by parallel sailing. Ocean navigation had created new business for chart makers, who interpreted mariners' sketches and logbooks, and drew inspiration (and coastlines) from one other's renderings. Not surprisingly, they found greater agreement in latitude than longitude, reconstructed all too often from vague accounts of flawed dead reckoning, which distorted continents and put capes and bays in the wrong place.

These distortions are apparent in the 1606 world map (fig. 3.6) of Dutch map publisher Willem Janszoon Blaeu, who sold world atlases as well as nautical charts.[40] Map historians credit Blaeu with actively seeking out up-to-date journals and encouraging mariners to report deficiencies of competing products,[41] an enterprising spirit reflected in his map's comparatively accurate trans-Atlantic distances. Along the less well-explored Pacific Rim, by contrast, Blaeu's coastlines exaggerate the thickness of lower South America and grossly overstate the longitudinal extent of North America. And closer to home,

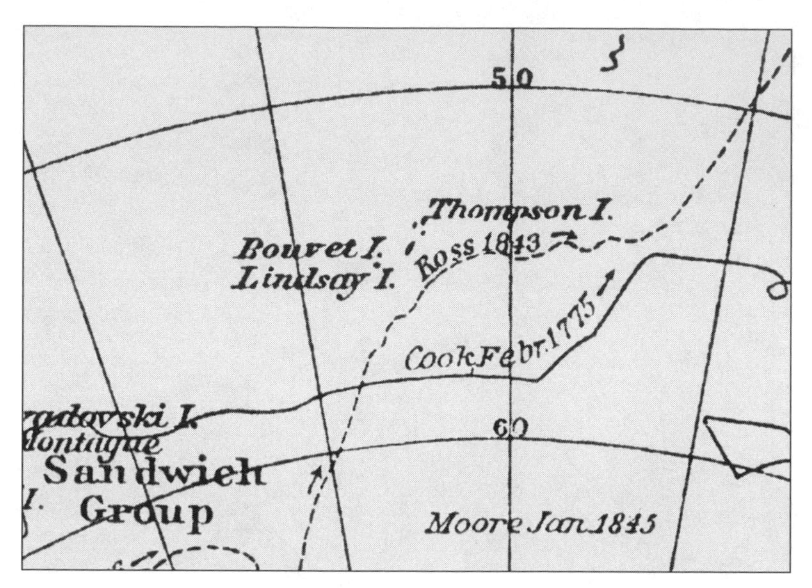

Figure 3.7. Excerpt showing Bouvet, Lindsay, and Thompson Islands, enlarged 300 percent, from "South Polar Regions," in *"The Times" Atlas,* 1896 edition, 113.

the Mediterranean Sea is 12 degrees of longitude too long.[42] Despite these flaws, cartographic scholar Günter Schilder credits "the map [with] a lasting influence on other printed and manuscript maps produced in the first half of the seventeenth century."[43]

However dependable for reaching destinations with correctly charted latitudes, parallel sailing kept ships at sea longer than necessary, eroding owners' profits and captains' purses, and increasing the likelihood of scurvy among poorly nourished sailors. In 1714 the British Parliament recognized the importance of longitude to navigation by offering £20,000 (over $15 million in today's money) for a dependable solution. As Dava Sobel describes so poignantly in *Longitude: The True Story of a Lone Genius Who Solved the Greatest Scientific Problem of His Time,* the money eventually went to John Harrison, a self-taught clockmaker who completed a working prototype in 1735 and spent the next four decades perfecting his chronometer and trying to claim the prize.[44] Although Harrison's invention was not adopted instantly or universally, it was influential in improving marine navigation and nautical charts.

Reliable longitude gave nineteenth-century world maps more credible coastlines, but inaccurate reports of small, insignificant islands cluttered the cartographic landscape well into the early twentieth century. In *Lost Islands: The Story of Islands That Have Vanished from Nautical Charts,* oceanographer Henry Stommel describes an 1865 British Admiralty chart covering the area between the equator and 30°S and stretching from 170°E east to 120°W.[45] The chart put Fiji and Tahiti in their right locations but displaced nine smaller islands, some to the east and some to the west, with errors in longitude between 6 and 15 degrees in contrast to disparities in latitude generally less than 2 degrees.

Equally bizarre is Bouvet Island, an uninhabited Norwegian possession more than 1,300 nautical miles south southwest of the Cape of Good Hope. According to Stommel, "almost any atlas up until 1930 shows a group of three islands, Lindsay, Bouvet, and Thompson, in the neighborhood of where today only Bouvet is shown."[46] Although a canvass of my university's map library suggests Stommel exaggerated the trio's persistence, I found them on a map of the "South Polar Regions" in an 1896 edition of the revered *Times Atlas* (fig. 3.7).[47] Thompson Island might have been a volcanic island that exploded in the late nineteenth century, while Lindsay Island represents the "rediscovery" of Bouvet Island in 1808 by Captain James Lindsay—a detail ignored when chart makers repositioned the original Bouvet Island, discovered in 1739 by Captain Pierre des Loziers Bouvet, who inaccurately reported its longitude.[48] Lindsay Island was voted off the map after an 1898 German expedition concluded Bouvet and Lindsay were one and the same, and Thompson was deleted from Admiralty charts after a 1930 British expedition failed to find it.

As this chapter underscored the difficulties of plotting the high-water line from offshore in the pre-chronometer era, the following chapter examines the substantially improved delineations resulting from trigonometric triangulation and systematic topographic and hydrographic surveys in the eighteenth century.

FOUR

TRIANGLES AND TOPOGRAPHY

By the nineteenth century, mariners could readily estimate longitude at sea, and chart makers no longer sought out logbooks to improve the depiction of generalized coastlines. Exploration had given way to commerce, and as the 1848 world sailing chart in figure 4.1 illustrates, marine cartographers had mastered the basic shapes of continents and moved on to winds, currents, and geographic advantage. Prepared by Matthew Fontaine Maury, a lieutenant in the U.S. Navy, the chart accompanied a congressional report on the emerging China trade.[1] Laid out on a Mercator projection, it shows representative routes and distances, in nautical miles, as well as the divide between coastal areas closer to an American port than to the country's chief rival, England.[2]

Appointed superintendent of the Navy's Depot of Charts and Instruments (later the Naval Observatory) in 1842, Maury believed strongly in systematic data collection. In an 1843 presentation to the National Academy for the Promotion of Science, he argued for furnishing all merchant ships with blank charts and requesting captains "to lay off the tracks of their vessels . . . every day, with remarks showing the time of year, the direction of the winds, the force and set of currents, and embracing, generally, all subjects that tend in any manner to illustrate the navigation of the seas through which they sail."[3] The following year the secretary of the Navy gave Maury permission to start compiling his Winds and Currents Charts,

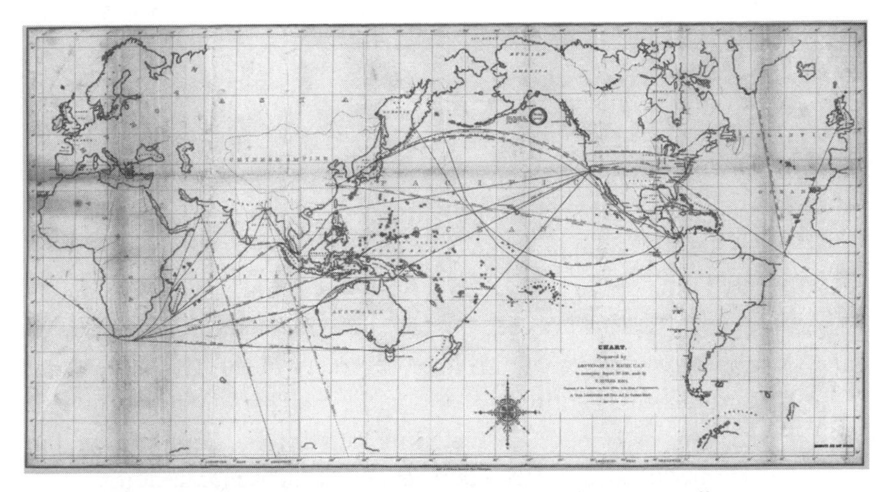

Figure 4.1. Chart prepared by Lieutenant M. F. Maury, U.S. Navy, to accompany the report by T. Butler King, Chairman of the House Committee on Naval Affairs, on steam communication with China and the Sandwich Islands, May 4, 1848. From the Library of Congress American Memory Web site, http://memory.loc.gov/ammem/.

widely appreciated by ship owners eager to save time.[4] Captains who turned over their notes received the charts free, and others could purchase a set. By the start of the Civil War in 1861, the Depot had given away or sold over 200,000 of Maury's charts.[5]

However useful in shortening mariners' journeys, small-scale sailing charts were less essential than detailed maps of coastal waterways and submerged dangers. Although private publishers in the United States and Europe were supplying the growing American merchant marine with coastal charts and pilot books (books of sailing directions),[6] they lacked the equipment and trained staff for a systematic survey of coastal waters. Thomas Jefferson, president from 1801 to 1809, recognized coastal charting as part of the government's responsibility to safeguard commerce and defend the country from attack, but the Survey of the Coast authorized by Congress in 1807 did not publish its first official charts until the mid 1840s.[7] While this delay partly reflects political bickering and the War of 1812, the geometric framework required for systematic coastal charting was a costly and complicated endeavor, difficult to expedite if done well.

The catalyst was Ferdinand Rudolph Hassler, a Swiss émigré who arrived in Philadelphia in 1805 with a collection of scientific books and precision instruments. Hassler promptly impressed the American Philosophical Society with his expertise in mathematics and geodesy (the science of measuring the earth), and in separate letters to the president, several members hailed his presence as a valuable opportunity.[8] France, Denmark, England, and Spain had established hydrographic offices in 1720, 1784, 1795, and 1800, respectively,[9] and the potential benefits to the United States, heavily dependent on coastwise shipping as well as international trade, were at least as great as in Europe. In February 1807 Congress appropriated $50,000 for the survey, and a month later Secretary of the Treasury Albert Gallatin asked scientifically qualified candidates to submit a proposal. Twelve responded, including Hassler. A committee of the American Philosophical Society reviewed their submissions, found Hassler's proposal the most detailed and convincing, and recommended his appointment as superintendent.[10]

Applying techniques refined in the eighteenth century by the Cassini family for its pioneering national survey of France,[11] Hassler's plan called for a well-orchestrated integration of geodetic triangulation, astronomical observation, topographic description, and hydrographic sounding. The key component was a geodetic survey based on a chain of triangles extended along the coast. Described schematically in figure 4.2, the network included lines of sight between control stations roughly thirty miles apart and two precisely measured baselines, about eight miles long on comparatively level land at far ends of the chain. At each control point a field observer would carefully measure the angles between converging lines of sight. Working outward from the baselines, an office assistant skilled in computation could then calculate a length for each of the network's other links using trigonometric tables describing constant, carefully computed numerical relationships among the sides and angles of triangles. Trigonometry was a godsend for the nineteenth-century geodesist because precise measurement was much easier for angles than for distances.

Trigonometry also promoted the efficient use of tedious, time-consuming astronomical estimates of latitude and longitude. To tie the geodetic network down to the planet's grid of meridian and parallels, Hassler called for astronomical observations at extreme

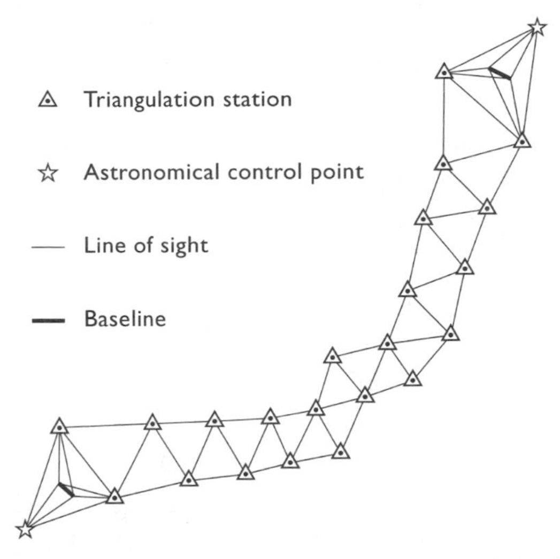

Figure 4.2. Key elements of a primary geodetic triangulation network. Smaller triangles nested within the primary triangulation and added along the periphery yielded a denser network of lower-order stations.

northeast and southwest ends of the network—in Maine and Louisiana (Texas was still part of Mexico). Prolonged observation of the heavens at these control points provided a precise fix on their spherical coordinates, which trigonometric calculations could then propagate throughout the network.[12] Because the earth was flattened somewhat at its poles, a pair of widely separated observatories promised more accurate results than a single astronomical station.

Trigonometry was equally useful in diffusing geodetic precision downward through a hierarchy of lower-order triangulation networks nested within or adjacent to the highly exact primary network. Multiple surveying parties working on different parts of the coast with less expensive equipment could develop a denser mesh of secondary and tertiary control points tied into an integrated national framework. Survey markers placed at convenient intervals along the coast could then provide a frame of reference for topographic surveyors, who mapped the shoreline and coastal features with accurate but less demanding instruments and in turn provided a geometric framework for hydrographic surveyors working offshore taking soundings for large-scale coastal charts.

Hassler described his strategy in a 189-page article published in

1825 in the *Transactions of the American Philosophical Society*.[13] He had submitted the manuscript five years earlier, but the society's lack of funds delayed publication.[14] Hassler himself was also needy. In 1818 Congress had transferred the Survey of the Coast from the Treasury Department to the Navy, which dismissed its civilian employees, including the superintendent, who managed to support his family in various occupations, including tutor, farmer, and land surveyor, as well as by writing textbooks and selling off family heirlooms.

Elimination of Hassler's job was one of many setbacks. Appointed to the post in 1807, he received no money for salary or instruments, but found work as a mathematics professor, first at West Point, which fired its civilian instructors in 1810, and then at Union College, in Schenectady, New York, where his eccentric behavior alienated students and colleagues. In August 1811, after Gallatin gave him $25,000 to purchase instruments, he embarked on a buying trip to Europe, where he was stranded until 1815, when the War of 1812 ended in a stalemate. Home at last, he organized an office, recruited field assistants, and initiated geodetic triangulation south of New York Bay in 1817. The Navy took over the following year and halted work on the primary network, which government leaders considered needlessly expensive. In offering a workable plan that challenged conventional wisdom, his *Transactions* article sought to stimulate support for a comprehensive coast survey that tied all control points to a common framework.[15]

Writing for men who valued systematic observation, Hassler stressed the importance of precision instruments, thoughtful planning, and rigorous measurement. A section titled "A Catalogue of the Instruments and Books collected for the Survey of the Coast" lists forty-nine separate items, the first of which is "One theodolite, of two feet diameter, made by Mr. Troughton."[16] An accompanying drawing by Hassler's oldest daughter, Carolyn, describes in detail its sighting telescope, the large graduated circular scale for reading horizontal angles, and the leveling screws necessary for a perfectly vertical axis (fig. 4.3). A reliable theodolite was the keystone of geodetic triangulation, and Hassler had "the highest expectations" for an instrument "executed under my own inspection by that distinguished artist Mr. Edward Troughton of London, agreeably to our united

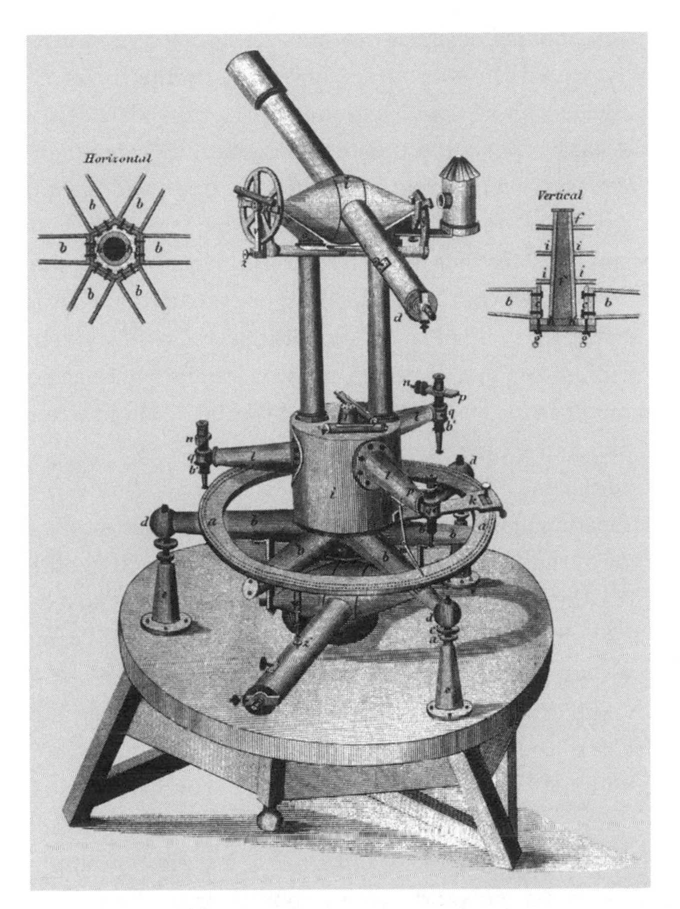

Figure 4.3. A perspective view of the Troughton theodolite, drawn by Carolyn Hassler and published in Hassler, "Papers on Various Subjects Connected with the Survey of the Coast of the United States," plate V. From the NOAA Photo Library, http://www.photolib.noaa.gov/library/images/big/libr0059.jpg.

views, and with that interest for its success, which the great friendship with which he was pleased to favour me could alone inspire."[17] Professional camaraderie was important to Hassler, who had moved next door to help out when the instrument maker was ill, and named his eighth child, born in 1813, Edward Troughton Hassler.[18]

Caroline provided other illustrations, and Hassler included a nine-page description of the large theodolite, a fourteen-page manual on its use and operation, and similarly detailed instructions for additional instruments, including two smaller, slightly less exact one-foot theodolites designed for measuring vertical and horizontal angles. Under the quirky section title "Promiscuous Remarks upon the Principles of Construction, the Choice and Trial of Instruments,"[19] Hassler pondered the idiosyncrasies of chronometers, the quality of glass used in manufacturing telescopes, and the improved "dividing engines" for inscribing the circular scales of theodolites and sextants.[20] Wary that minutiae should never become the weakest link, he warned that "the [packing] boxes ought always to be of light and straight grained wood," and that no part of an instrument "must ever be fixed in a direction diagonal to the grain, as it will be pressed out of shape by the drying of the wood."[21] No source of error was too small to escape comment.

Toward the end of his article, in a section titled "On the Mechanical Organisation of a Large Survey, and the Particular Application to the Survey of the Coast," Hassler addressed the selection of triangulation sites, the need to transport instruments with utmost care and shelter them from sun and weather, and other practical concerns.[22] Finding elevated locations with long, unobstructed lines of sight to other triangulation stations required careful reconnaissance. Stations that are "the highest and freest in the neighbourhood," he observed, "will not only accelerate the work . . . but also enable the operator to choose the most advantageous combination of triangles."[23] The perfect station was free of tall trees and on firm ground so that an observer could move around without disturbing a carefully leveled instrument. Less ideal situations forced survey parties to cut timber, clear brush, and construct towers or elevated platforms.

Hassler's instructions for measuring a baseline are daunting. "In all surveys of considerable extent," he cautioned, "the exact determination of the line, which forms the base of the whole triangulation, is of the greatest importance." In addition to being "the most tedious part of the work," running an accurate baseline "presents, in its mechanical execution, difficulties which have always called forth . . . inventive genius."[24] Because geodetic surveyors measured distance by laying precisely machined iron bars end to end, they required a perfectly straight path, typically contrived by fitting together a series of long, narrow wooden trays, open at the ends. Supported by tripods and carefully leveled, the trays provided an extendable plat-

Figure 4.4. Coast Survey crew measuring a base line in Epping Plains, Maine, 1857. From the Historic Coast and Geodetic Survey Collection, NOAA Library, http://www.photolib.noaa.gov/historic/c&gs/images/big/theb1646.jpg.

form on which four bars, each two meters long, could be leapfrogged forward.[25] At each step workers lifted the bar at the rear, carried it forward, and placed it carefully at the front while advancing the trays in a similar fashion, to receive the next increment.

I couldn't find a picture of Hassler's apparatus, but figure 4.4, an 1857 engraving from the NOAA photo archives, depicts a tedious task requiring rigorous monitoring. In addition to sheltering the advancing chain of bars from weather and direct sunlight, survey crews checked the alignment of the sections with a "directing telescope," measured the bars' temperature with a thermometer, and calculated an adjustment for thermal expansion.[26] Thanks to trigonometric triangulation, few distances needed direct measurement.

Because the national network that evolved later in the nineteenth century needed many more baselines than the single pair in Hassler's plan, geodetic surveyors were constantly seeking more efficient and reliable ways to measure distance. Wooden sticks, which were too sensitive to humidity, had preceded metal bars, which gave way around 1845 to bimetallic rods encased in wood, which were not only less sensitive to temperature change but easier

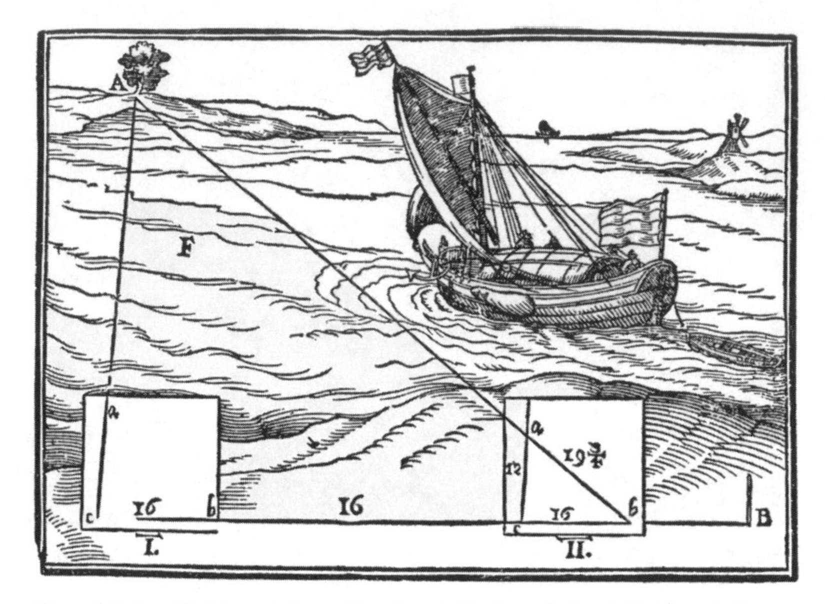

Figure 4.5. Graphic triangulation with a plane table. From Wainwright, *Plane Table Manual*, 3.

to manipulate—their casings replaced the wooden trays, as a close look at the engraving (fig. 4.4) will confirm.[27] More significant was the adoption around 1900 of thin tapes or wires about 300 feet long made of invar, a highly stable nickel-steel alloy. Much longer than the encased bars, invar tapes could be used on sloping or irregular terrain.[28]

Far less exacting was the plane table, which Hassler considered "the best method of surveying the minute details which are to fill up a triangulation."[29] An instruction manual issued a century later provides a more lucid description as well as an intriguing "early illustration" of unspecified origin, possibly from Holland—note the windmill (fig. 4.5). The sketch shows a triangle with points *B* and *C* a known distance apart and point *A* on the far side of a bay or harbor. "In its earliest form," we're told, the plane table "consisted of nothing more than a sketching board with a sheet of drawing paper fastened on one face, and an ordinary wooden ruler with a needle at each end as a convenient means for sighting one edge of the ruler toward a distant object."[30] In this example, the surveyor begins by anchoring the board in a level position at *C* and marking its loca-

Figure 4.6. The Coast Survey plane table. From Wilson, *Topographic, Trigonometric and Geodetic Surveying*, 153.

tion on the paper as point *c*. Using the sighting ruler, he draws lines from *c* toward objects *A* and *B*. After moving the plane table to *B*, he marks its position *b* along the line toward object *C* so that length *cb* represents the measured distance (16) between *B* and *C*. A third line, drawn from *b* in the direction of *A*, completes the map, which then can be measured with a ruler to find the distances from *A* to *C* (12) and *A* to *B* (19¾).

Plane tables used for coastal surveying included a tripod with a ball and socket mechanism for leveling the drawing board and an alidade with a telescopic sight attached to a straightedge aligned in the same direction (fig. 4.6).[31] When the telescope was pointed toward a graduated stick, or stadia rod, held vertically by an assistant several hundred feet away, a topographer could estimate distance by counting the number of divisions on the rod between short horizontal lines called stadia hairs ground into the top and bottom of the glass—the greater the number of divisions between the stadia hairs, the greater the distance from the plane table to the rod. With the plane table positioned for an advantageous view of the shoreline, the topographer watched the "rod man," who walked along the apparent high-water line, pausing to hold the rod vertically at each bend while the observer marked the rod's position on the drawing at the corresponding distance along the line of sight. After threading a smooth line through the dots, the topographer moved the plane table to another position along the coast, leveled the drawing board with a spirit level, fastened a new sheet of paper to the board, and used triangulation to tie into the local survey network. At each position he used further sightings on the stadia rod to sketch in roads, prominent buildings, fence lines, and other landscape features. Joined together, these field sheets yielded an integrated map of shoreline topography.[32]

After the topographic party mapped the landward side of the coastline, hydrographers in a sounding boat completed the survey by probing the water's depth with a pole or a "lead line," drawn downward by a twelve-pound lead weight.[33] To record location, the boat crew carried a map called a boat sheet, on which a surveyor had plotted the high-water line and marked the positions of tall, highly visible signals set up at known locations along the shore, and sometimes supplemented by signal buoys, tied into the topographic survey by triangulation. A member of the hydrographic party measured the angles between three successive signals (A, B, and C in fig. 4.7, *left*) with a sextant, while another observer plotted the boat's position with a three-armed protractor called a station pointer.[34] The center arm was anchored to a disc, graduated in degrees so that the two outer arms could be set to the angles measured with the sextant. When aligned with its arms extended through the marked positions of the shoreline signals (*a*, *b*, and *c* in fig. 4.7, *right*), the station pointer fixed the sounding at the point of convergence (*x*).

Inshore hydrography required additional probing to enrich the chart with the low-water line and bathymetric contours at depths six, twelve, eighteen, and thirty feet. While these enhancements demanded an accurately surveyed high-water shoreline, tied into the national network with plane table and stadia rod, crude graphic

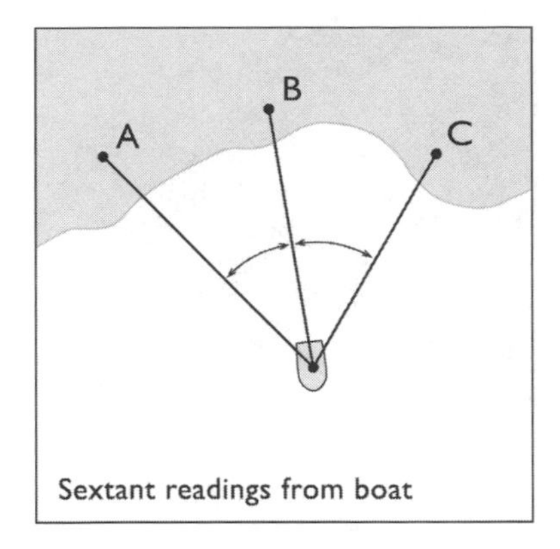

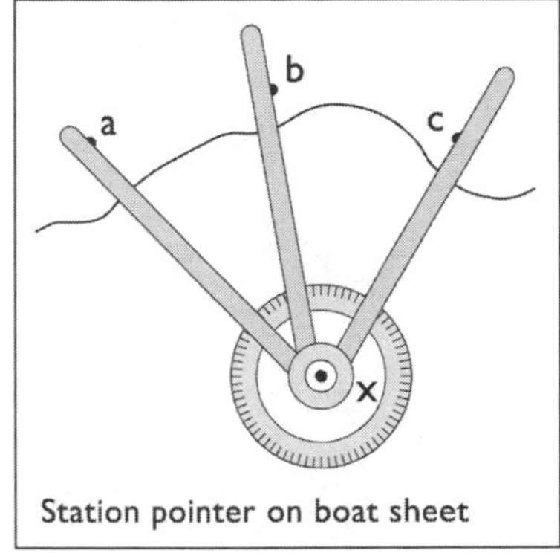

Figure 4.7. Angles between lines of sight to onshore signals (*left*) were replicated on the boat sheet with a station pointer (*right*), which located the sounding at the point of convergence.

triangulation with sextant and station pointer was fully adequate offshore, where the positional accuracy of soundings was less important than their bathymetric accuracy, based on a systematic canvass of the seabed.[35] Working outward from the shore along parallel sounding lines perpendicular to the coastline and about 500 feet apart, or along lines running due north or south across harbors or small bays, hydrographic parties took many more soundings than their nautical charts could show. Figure 4.8, which juxtaposes excerpts of a sounding survey and finished harbor chart for Mount Desert Island in eastern Maine, suggests that unpublished soundings typically outnumbered their published counterparts by more than twenty to one. Because mariners were wary of running aground, the single depth chosen to represent multiple soundings on the published chart was a minimum, not an average. And because chart users appreciated labels like "sft" and "rky," which rated the relative consequences of running aground on a soft or rocky seabed, coastal hydrographers sampled the seabed's composition as well as its depth.[36]

Although Hassler's 1825 plan said little about plotting the shoreline and taking soundings, his discussion of baselines and trigonometric triangulation impressed the country's fledgling scientific community and its friends in Congress, who in 1832, after repeated pleas, convinced their colleagues to return the Survey of the Coast

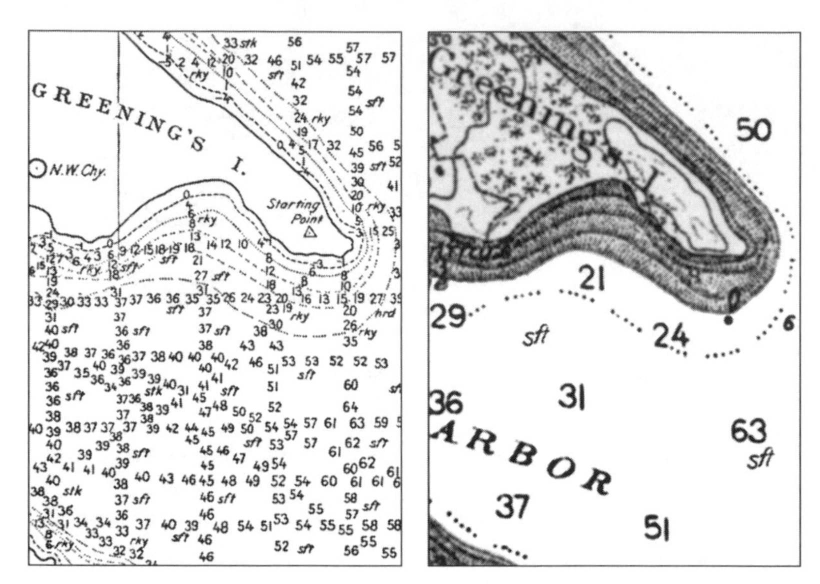

Figure 4.8. Excerpt from an early twentieth-century hydrographic survey (*left*) and corresponding portion of a completed nautical chart (*right*) for waters near Mount Desert Island, Maine. The finished chart, usually produced at a smaller scale than the hydrographic survey, is enlarged here for ready comparison. From Jones, *Elements of Chart Making,* figs. 22 and 31.

to the Treasury Department, which reinstalled Hassler as superintendent. Reactivation was delayed because several crucial instruments had been lost by the Navy, irreparably damaged, or loaned to Massachusetts, which was reluctant to return them.[37] The prized two-foot Troughton theodolite had to be sent back to England for repair, but Hassler seized the opportunity to order a larger, thirty-inch model as well as his own dividing engine "to enable us to have the instruments as well executed in this country as abroad, with much less time and more appropriated to our individual uses."[38]

Eager to resume triangulation, Hassler recruited military and civilian assistants, recovered control points he had marked in rock back in 1817, measured a 14,058.9850-meter baseline on Fire Island, off Long Island, and extended the network from New Jersey eastward into Connecticut and Rhode Island and southward into Delaware.[39] He remained in command after the quixotic president Andrew Jackson reassigned the agency to the Navy in 1834, and two years later it was back in the Treasury Department under a new name, the U.S. Coast Survey.[40] A focus on New York and its environs, including Newark Bay and Long Island, yielded several black-and-white sketches and preliminary charts in the late 1830s.[41] Hassler planned a growing series of elegantly lithographed nautical charts, reproduced on the Coast Survey's own printing press, delivered in 1842, from copperplates inscribed in its own map engraving department.[42]

The first official U.S. Coast Survey chart, a 1:30,000 portrait of New York Bay and Harbor, appeared in 1844,[43] shortly after Hassler's death in November 1843 from a severe respiratory illness brought on by prolonged exposure following a fall onto a sharp rock while he was trying to protect his instruments during a severe storm.[44] He was an odd bird, eccentric in his precisely enunciated, nonidiomatic English; his preference for woolen socks and white flannel in both hot and cold weather; and his insistence in making all major observations personally.[45] A control freak in matters of measurement, Hassler was fully committed, his biographer Florian Cajori observed, "to resist compromise, to oppose suggestions of alterations, made by engineers and statesmen, [and] to maintain this opposition against the adoption of 'cheaper' yet 'just as good' plans."[46] Despite his unconventional behavior, Hassler was respected by his subordinates, with whom he pioneered the modern government science agency.

Hassler's successor, Alexander Dallas Bache, had an equally impressive resume: great-grandson of Benjamin Franklin, West Point graduate, professor of natural philosophy and chemistry at the University of Pennsylvania, and superintendent of Philadelphia's public schools. Unlike Hassler, who was arrogant, forgetful, and not a people person, Bache was personable, administratively competent, and organizationally savvy. In contrast to Hassler's focus on the New York region, which he could supervise directly, Bache divided the coast into nine sections, for simultaneous work on baselines, triangulation, topography, and hydrography.[47] The new strategy required cordial relations with Congress, which bankrolled an increase in Coast Survey employment from 216 in June 1851 to 776 in February 1855. By 1858, this expanded workforce had completed topographic and hydrographic surveys for two-fifths of the nation's coastline, including the Pacific coast, and published more than fifty finished and eighty preliminary charts. A clear benefit, according to an 1858 study by the American Association for the Advancement of Science, was fewer shipwrecks, which saved the economy almost $3 million a year.[48]

Bache's greatest challenge was the Civil War (1861–65), which threatened to strip the Coast Survey of funding and personnel. A Union loyalist as well as an active member of Washington scientific community, Bache knew that the military needed scientific expertise as well as detailed coastal charts. He blocked attempts of Confederate sympathizers to abscond with maps and data, briefed Union officers with insightful interpretations of coastal charts, and proposed the Blockade Board, a small, top-secret organization that developed a strategic plan for enforcing the blockade of Confederate ports, recommended places to attack, and contributed directly to Union victories in several battles, including the taking of New Orleans.[49] The Blockade Board met regularly early in the war, worked long hours, and developed detailed guides, with accompanying charts, for the blockade squadrons. The description of Oregon Inlet, on the Outer Banks, illustrates the board's explicit advice: "Oregon, or New Inlet, 35 miles north of Cape Hatteras, Pamlico Sound, coast of North Carolina, has a dangerous shifting bar, on which there is 7 feet at low and 9 feet at high water. It should be blocked up by sinking as many hulks as necessary for the purpose. The place for these hulks will be shown on the chart."[50] Although Bache's grueling schedule as a military advisor hastened his death in 1867, he held the Coast Survey together by demonstrating the strategic value of detailed charting.

To prevent this chapter from swelling to several times its present length, and blurring my intended focus on mapping coastlines, I have ignored numerous significant scientific and technological accomplishments of Hassler, Bache, and their successors, most notably the Coast Survey's contributions to the theory and application of geodetic astronomy, terrestrial magnetism, gravity anomalies, and what geodesists call the Figure of the Earth.[51] Because reliable maps of the shoreline, coastal waters, and anything else worth mapping with precision require an understanding of the compass needle's deviation from true north, the plumb line's deflection from a direct line to the planet's center, and Earth's departure from a perfectly spherical shape, the agency responsible for charting the nation's coastline necessarily expanded its mission into geodesy and geophysics, and extended its reach far inland, most prominently with an arc of triangulation that followed the 39th parallel westward from Cape May, New Jersey, to Point Arena, California, to join the Atlantic and Pacific coasts into a single, unified geodetic network.[52] Initiated in 1871 and completed in 1898, the Transcontinental Triangulation, was one of many extensions of a primary geodetic network that grew in length from 7,000 miles in 1899 to 31,500 miles in 1932.[53] In 1878, Congress recognized the bureau's increased role by expanding its name to U.S. Coast and Geodetic Survey.

FIVE

OVERHEAD IMAGING

Five or six years from now I'll be both proud and relieved when *Cartography in the Twentieth Century,* the million-word encyclopedia I've been editing for over a decade, is finally in print as part of the *History of Cartography,* a massive, multivolume reference work for scholars. I mention this because the planning required me to identify fundamental modes of mapping and map use.[1] Most were carryovers from earlier eras, but one of the two modes that emerged since 1900 is overhead imaging, which had a huge impact on coastal cartography.[2] As this chapter reveals, airborne sensors assumed a key role in delineating the land-water boundary, now represented by contrasting tones and textures as well as traditional, graphically crisp shoreline symbols.

Few technological innovations spring up unheralded, so it's not surprising to find overhead imaging conspicuously rooted in the nineteenth century. In 1827 Joseph Nicéphore Niépce, a French lithographer, took the first photograph, and three decades later, in 1858, Gaspard-Félix Tournachon, a portrait photographer looking down on Paris from a hot-air balloon, took the first aerial photo.[3] During the Civil War, the Union Army used reconnaissance photos shot from tethered balloons to spy on Confederate positions in eastern Virginia but phased out its balloon corps after two years. Despite a few intriguing demonstrations in Europe and America, air photos taken from balloons or kites had little impact on mapping until the early 1900s, when the self-propelled balloon, or dirigible, provided a maneuverable, comparatively stable platform.

An otherwise comprehensive 1912 textbook on topographic and geodetic mapping by Herbert Wilson, chief engineer at the U.S. Bureau of Mines, epitomizes the slow acceptance of overhead imaging. Obviously aware of photography's value as a measurement tool, Wilson completely ignored aerial platforms. His single chapter on "photographic surveying" focuses on measuring angles and distances using photographs shot from high peaks in places like Alaska and northwestern Canada, with great relief and few trees.[4] Anchored on the ground and carefully leveled, a photo-theodolite could capture geometric relationships and topographic details quickly during brief periods of sunshine, thus transferring much of the effort from the field to the office. Choosing his words carefully, Wilson opined "that a fair map can be made by photo-topographic methods, under favorable conditions, more rapidly in the field and at less cost than a good map can be made on the same scale by plane-table methods."[5] Impressed with the uncharted vastness of northern North America, he recognized efficiency when he saw it.

Although Wilson failed to anticipate the airplane's rapid evolution during World War I and its decisive displacement of the dirigible as the best way to position a camera over the right spot at the right time,[6] his distinction between "a fair map" and "a good map" resonated three decades later with Coast and Geodetic Survey mapping guru Aaron Shalowitz. Writing in the October 1945 issue of *Scientific Monthly,* Shalowitz compared plane tabling with aerial surveying. "For mapping from the ground," he asserted, "the planetable is still the most satisfactory instrument." Even so, "ground topographic methods are rapidly giving way to the more economical and more expeditious method of aerial topography [because] the wealth of information and fullness of detail embraced in an air photographic survey cannot be matched by any other practicable method of surveying."[7] Two decades later, in his landmark *Shore and Sea Boundaries,* Shalowitz not only declared aerial photogrammetry "the greatest advance in topographic mapping since the prototype of the modern planetable was developed . . . in the latter part of the 16th century," but credited air surveys with having "further increased the accuracy of the chart, particularly in inaccessible areas [and bringing] within practical scope the immediate revision of areas where natural or man-made changes have occurred—an important factor in safeguarding the sealanes."[8]

As Shalowitz implied, coastal cartography's adoption of aerial surveying reflects a significant cost savings as well as the comparative ease of photographing vast or remote areas quickly. No less important was the rapid development of civil aviation after World War I, and pilots' growing demand for up-to-date maps of the shoreline and other prominent ground features. In 1926 the Air Commerce Act made the U.S. Coast and Geodetic Survey responsible for mapping airways, which could be tied into its national triangulation network. Geodetic surveying provided a framework for topographic mapping, after all, and the Survey, which started experimenting with aerial photography in 1919, had maintained a photogrammetry unit since 1922.[9]

Experimental work focusing on Florida included an air-photo resurvey carried out around 1929 along the Gulf Coast near Naples.[10] Juxtaposition of before-and-after nautical charts (fig. 5.1) illustrates the increased topographic detail that cost no more than a conventional ground survey confined to a very narrow strip of land along the shoreline.[11] As the "after" map shows, the aerial survey added roads, railway tracks, and the shoreline of Naples Bay—useful detail for pilots trying to plan or follow a course, and especially valuable for city planners, highway departments, and officials in charge of inland waterways. Eager to promote photogrammetric mapping, the agency's director unabashedly confessed in his annual report for 1930 that "for several years lack of adequate personnel and funds has compelled the bureau to abandon its long-established practice of engaging regularly in revision of topographic surveys of the coasts."[12] Aerial imagery would put chart maintenance back on course.

Another benefit was the air photo's unassailable confirmation of dramatic change. As pioneer Coast and Geodetic Survey photogrammetrist Oliver Reading remarked in a 1931 assessment, "In the study of shoreline change, the faithfulness and dependability of the representation of detail is a great advantage, for changes are frequently so radical as to throw doubt on the accuracy of the surveys."[13]

Continuing a tradition of creating new technology to meet unique

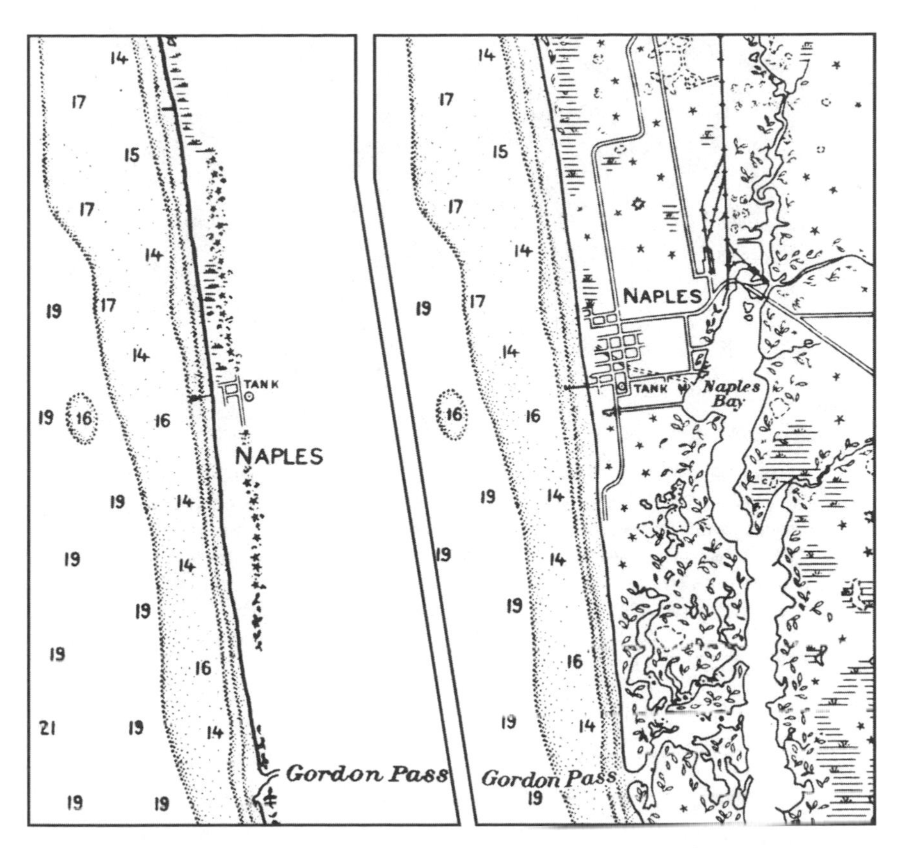

Figure 5.1. Comparison of nautical charts for the Florida coast near Naples before and after an air-photo resurvey, ca. 1929. Excerpt from a larger illustration in U.S. Coast and Geodetic Survey, *Annual Report of the Director, United States Coast and Geodetic Survey to the Secretary of Commerce for the Fiscal Year Ended June 30, 1930*, opposite 18.

needs, Coast and Geodetic Survey engineers designed a 300-pound, nine-lens aerial camera that could photograph an eleven-by-eleven-mile area in a single shot.[14] Each lens cast a separate image on a shared piece of film 23 inches across (fig. 5.2, *left*). At the center was a vertical view, captured with a lens aimed downward at a point on the ground called the nadir point, directly beneath the aircraft. Surrounding the center image's octagonal border were eight oblique views, each tilted outward 38 degrees from the vertical in directions 45 degrees apart. Scale, which is relatively consistent on a vertical photo taken from an altitude of a mile or more, decreases markedly toward the far edge of an oblique photo. The nine-lens camera

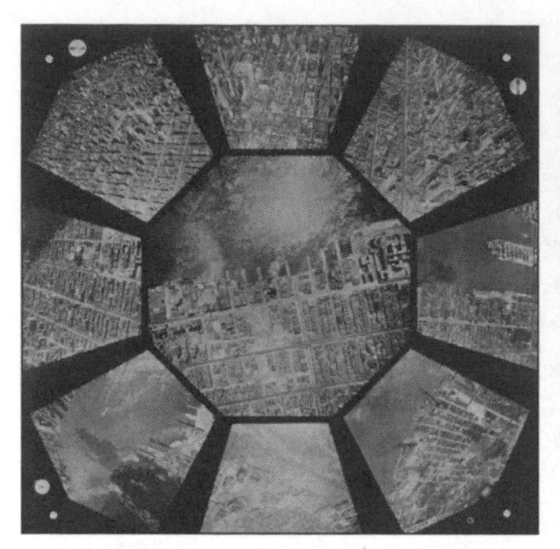

Figure 5.2. A "transforming printer" converted the single vertical and eight oblique images (*left*) captured simultaneously by the Coast and Geodetic Survey's nine-lens camera into a single composite image with the geometric properties of a vertical photo (*right*). Taken over New York City at a lower than normal altitude, these images illustrate the printer's prowess in reconstructing Manhattan's rectilinear street grid, noticeably distorted in the oblique views. From the NOAA Photo Library, http://www.photolib.noaa.gov/historic/c&gs/images/big/theb0944.jpg (*left*) and http://www.photolib.noaa.gov/historic/c&gs/images/big/theb0943.jpg (*right*).

would have been far less useful without the ingenious transforming printer that recast all nine images onto a single, composite print (fig. 5.2, *right*), 35 inches square and similar in appearance to a normal vertical air photo. And if the central camera were not pointed directly downward at the time of exposure, a rectifying camera could remove the resulting distortion, called tilt, as well as produce a new, truly vertical image at a scale consistent with air photos for other parts of the coast.

With the nine-lens camera, an airplane flying at 14,000 feet could capture 1:20,000 imagery for an eleven-mile-wide strip—suitably detailed for coastal mapping and sufficiently broad to include onshore topography and offshore islands. By contrast, conventional, single-lens aerial photography, acquired in relatively narrow parallel strips, needed many more individual photos to cover the same area. Shot from varying altitudes with the camera's optic axis tilting slightly in diverse directions by different amounts, these individual photos were not easily combined into an equivalent composite image—removing their geometric inconsistencies was a labor-intensive process requiring sophisticated photogrammetric instruments and experienced technicians.[15] What's more, the nine-lens camera offered comparable accuracy with only a quarter as many ground control points, the name for the precisely surveyed landmarks and field targets used to tie air photos into the triangulation network.[16] The large format of the nine-lens camera was inherently advantageous because ground control accounted for half the cost of making topographic maps from aerial photography.[17]

As the Manhattan skyscrapers in the transformed vertical composite illustrate (fig. 5.2, *right*), an air photo is a perspective view, not a map. Lines of sight converge toward the lens, and tall buildings appear to lean away from the center of the photo, which makes some people queasy, as if they were looking out a gaping hole in the bottom of an airplane. A larger, more detailed print might show the top corner of a tall building displaced outward from the corresponding bottom corner, and a vertical flagpole, if you could find one, would also appear to lean. All displacement is along lines radiating from the photo's center (if it's truly vertical), and the degree of displacement increases with distance from the center and difference in elevation.

An annoyance to anyone eager to use an air photo as a map, radial displacement becomes an asset when a pair of overlapping photos is viewed under a stereoscope, which affords a dramatic three-dimensional view of hills, valleys, skyscrapers, and other terrain features. And when properly mounted on a specially designed stereo plotter, this visual model of the terrain lets the photogrammetrist trace contour lines, measure differences in elevation, and accurately position buildings, roads, and other features in the horizontal plane represented by the map. During the 1930s and 1940s, aerial photogrammetry ousted the plane table as the principal technique for topographic mapping,[18] and four decades later computer software able to "see stereo" automatically generated displacement-free images called orthophotos (fig. 1.6, *left*).[19]

Less sophisticated technology was convenient in coastal mapping, where low relief minimizes radial displacement and cartographers can trace the land-water boundary on tide-coordinated photos viewed in a projector designed to remove tilt.[20] With a simple pocket stereoscope like those sold in army surplus stores in the late 1940s

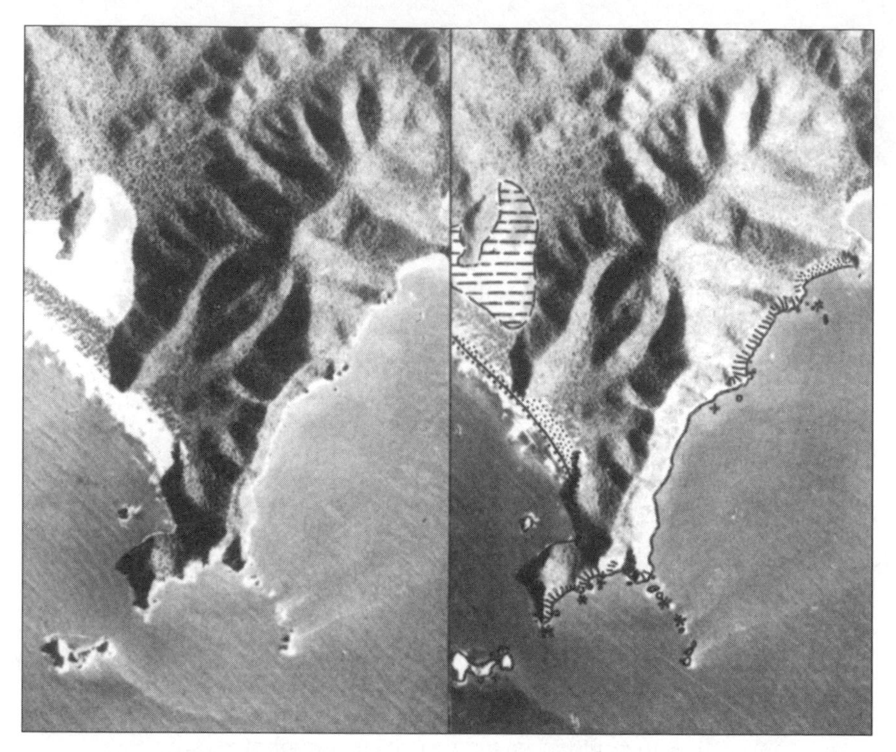

Figure 5.3. Overlapping U.S. Navy images of partially submerged mountainous terrain, positioned for stereo viewing. Sixty-percent overlap between successive exposures provides three-dimensional viewing along a strip of air photos. From McCurdy, *Manual of Coastal Delineation from Aerial Photographs*, 95.

for a few dollars, a mapmaker could identify hazards and delineate noteworthy differences in vegetation and surface material.[21] Annotations to the right half of the Navy Hydrographic Office imagery in figure 5.3 illustrate the details a savvy photo interpreter could glean from overlapping air photos. Telltale tones and textures viewed in stereo reveal not only the high-water line but also rocks awash (asterisks), sunken rocks (plus-signs), sandy beaches (stippled pattern), coastal cliffs undercut by waves (short strokes perpendicular to the shoreline), and a large, periodically inundated area (horizontal dashed lines) just behind the beach.

As this example suggests, air photos can help cartographers trained in coastal geomorphology understand the physical processes affecting the shoreline and identify areas prone to erosion or deposition. Reconnaissance photography, taken every one to three years with a light plane flying at low altitude along changeable coastlines, could be used without ground control for making minor revisions to otherwise accurate charts as well as for detecting areas requiring remapping or extensive revision. If comparison with existing charts revealed only modest change, minor discrepancies could be corrected without a more complex and costly resurvey.[22] Aerial photography also helped mapmakers detect changes to roads, piers, and other man-made features.

While air photos are handy for making minor corrections, unless you're planning an amphibious assault on hostile shores, a more thorough revision is not as easy as the annotated Navy photos in figure 5.3 might imply. The snag for domestic nautical charting was the official, mean high-water shoreline, which required not only tide-coordinated imagery but also a comprehensive "shoreline inspection" carried out prior to stereo compilation. Unlike roads and other inshore topographic features, which needed little supplementary fieldwork, the legal shoreline demanded direct observation of debris and other beach characteristics, carefully marked on a set of photos for use back at the office. Although the field inspector could cover much more ground in a day than a survey party with a plane table and stadia rod, he had to walk the shoreline looking for evidence of tidal action as well as rocks and other hazards, above or below the low-water line. Responsible for important details not apparent on the photos alone, the inspector also recorded bridge clearances, the names and configuration of piers, and other features that make the coastline more than just a line on a map.[23]

New materials and improved equipment reinforced the air photo's pivotal role in coastal cartography. In 1958 the Coast and Geodetic Survey switched from black-and-white to color photography, for more revealing depictions of land and shallow water.[24] Three years earlier the superior land-water contrast of black-and-white infrared emulsions (fig. 1.6, *right*) had provided a more efficient way to identity the mean lower-low-water shoreline on tide-coordinated photography, and in 1969 the agency adopted color-infrared photography for special projects concerned with currents, seafloor habitats, and land cover. In the 1970s electronic scanners in orbit hundreds of miles above the surface started offering similar color, infrared, and color-infrared imagery, but spaceborne remote sensing—except for whatever technology the intelligence community might have—has yet to match the resolution and flexibility of its airborne counterpart.

Although the role of satellite platforms in coastal cartography is still largely experimental, the electronic scanning of conventional photography fostered a dramatic advance in stereo compilation in the 1990s, when digital photogrammetry supplanted analytical photogrammetry, an interim computer-assisted technology that had displaced analog photogrammetry in the 1960s and 1970s.[25] Analytical photogrammetry offered more systematic control of error than its optical-mechanical parent, and digital photogrammetry's sophisticated software gave less experienced technicians using desktop computers a better, cheaper, faster way to make maps from overhead imagery. The global positioning system (GPS) further aided the computerized reconstruction of stereo geometry by tagging every image with the camera's elevation, latitude, and longitude at time of exposure. And somewhere during the transition from instrumental to digital photogrammetry, walking the high-water line became unnecessary. NOAA still field checks its charts, but the Hydrographic Surveys Division, which is responsible for the inspection, works mostly from the water, usually seaward of the four-meter isobath to avoid rocks and hidden hazards.[26]

NOAA's Remote Sensing Division—the Photogrammetry Division was renamed in 1997,[27] about the time cartography became geographic information science—is responsible for a variety of experimental projects as well as operational photogrammetric mapping. Imaging technologies under investigation have narrowly defined applications. Hyperspectral remote sensing, for instance, splits the electromagnetic spectrum into very narrow wavelength intervals, called bands, some of which are especially powerful for identifying specific types of ground cover, such as railway bridges and airport runways.[28] Synthetic aperture radar (SAR) and interferometric synthetic aperture radar (IFSAR), both of which can penetrate dense cloud cover, even at night, have proved useful for mapping remote shorelines in Antarctica and Alaska.[29] NOAA's Coast and Shoreline

Change Analysis Program (CSCAP) overlays high-resolution satellite data on existing charts of the country's larger ports to detect changes in the man-made shoreline. Charts can be updated immediately to show minor changes or scheduled for revision using standard airborne photogrammetry, which has the requisite resolution and accuracy.[30]

Most promising among these new technologies is light detection and ranging, better known as lidar.[31] Similar in principle to radar and sonar, which probe the environment with their own signals, lidar estimates distance by measuring the precise time a pulse of laser light takes to reach a target and return to the sensor—the longer the time, the greater the distance. Carried aloft by an airplane, the sensor pings the surface below with a pulsating laser beam it swings rapidly back and forth within a swath perhaps 650 meters (0.4 miles) wide as the plane flies along the beach.[32] Each hit yields a point on the surface below. Its latitude, longitude, and elevation are calculated from the direction of the beam at the time of transmission, the return time of the "backscattered" (reflected) light, and the aircraft's instantaneous position, determined by onboard GPS and inertial-navigation units. Although specifications vary with flying speed, flight height, and the lidar sensor itself, points are typically 1.5 meters (5 feet) apart horizontally and elevations are accurate to within 15 centimeters (6 inches).

Because the sensor fires individual pulses at a rapid rate, the mission yields a massive, dense point cloud of elevation points, which must be registered to a common reference surface—mapmakers call this the horizontal datum—and then filtered to remove buildings, vegetation, and anything else projecting above the bare-earth surface shown on most topographic maps. Transforming the tens of millions of elevation points in a typical lidar mission is a computationally demanding process, made even more burdensome by the correction for backscatter from roofs and the tree canopy.[33] The before-and-after shaded-relief maps in figure 5.4 demonstrate the effectiveness of bare-earth filtering for a narrow portion of North Carolina's Outer Banks.[34] As shown in figure 5.5, the formidable challenge posed by structures and vegetation cover attests to the need for filtering and the algorithm's dexterity.

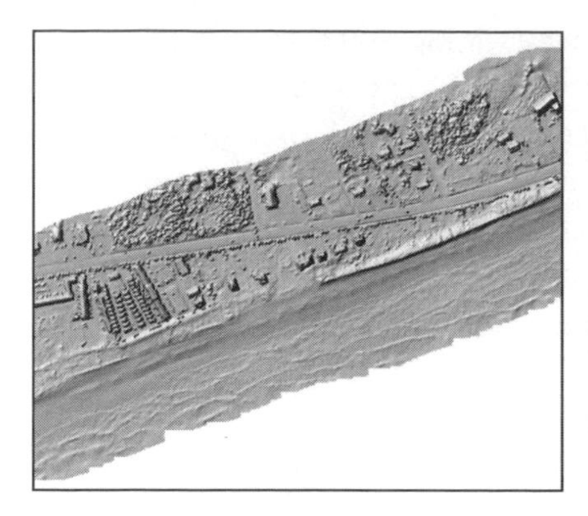

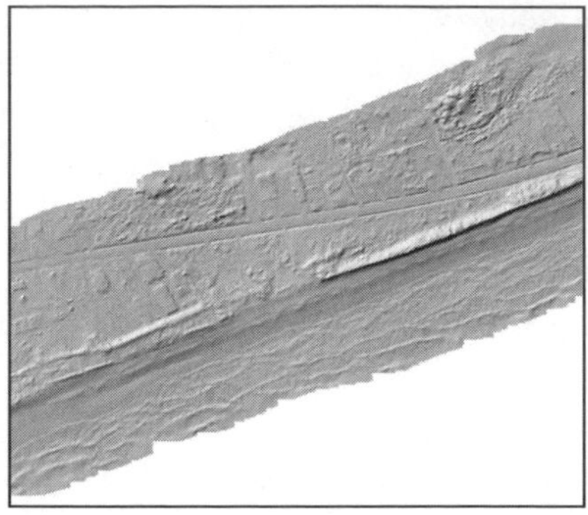

Figure 5.4. Shaded-relief maps for a barrier island near Cape Hatteras compare first-surface lidar topography (*left*) with the same scene stripped of buildings and vegetation by bare-earth filtering (*right*). The dark line on both images is the shadow of a sand dune parallel to the shoreline. Courtesy of Keqi Zhang and Dean Whitman, International Hurricane Research Center at Florida International University.

Figure 5.5. Overhead image of the area in figure 5.4. Courtesy of Keqi Zhang, International Hurricane Research Center at Florida International University.

Lidar is especially advantageous with sensors designed to collect simultaneous datasets for near-shore topography and shallow-water bathymetry. A pioneer was the SHOALS (Scanning Hydrographic Operational Airborne Lidar Survey) program, initiated by the U.S. Army Corps of Engineers in the early 1990s. A specially configured lidar scanner probes shallow offshore waters using laser pulses with two wavelengths, red and green.[35] Over water the red light bounces upward from the surface while the green light penetrates as much as 50 meters (160 feet) to reflect from the seafloor. Separate elevation points for the red and green pulses indicate the surface and bottom of the water, respectively. Once the computer generates separate, continuous surfaces for the top and bottom of the water, simple subtraction yields the depth.

Bathymetric lidar, also called topo-bathy lidar, holds the promise of a "seamless" topographic/bathymetric dataset, which the National Research Council considers essential "if coastal zone management is . . . to produce accurate maps and charts so that objects and processes can be seamlessly tracked across the land-water interface."[36] Seamless tracking is especially important for modeling coastal erosion and assessing vulnerability to storm surge, which is directed onto the shore by winds and the configuration of the seafloor. A seamless topographic/bathymetric dataset could also be integrated with tidal data to generate reliable shorelines for various legally significant tidal datums, including mean high water, mean low water, and mean lower-low water.[37] The promise of lidar includes a more precise way of delineating specific shorelines as well as a more comprehensive means of describing the changing coastline.

SIX

ELECTRONIC CHARTS AND PRECISE POSITIONING

Marked by a black line and highlighted with contrasting tan and blue tints for land and water, the high-water shoreline is the nautical chart's most prominent, if not its most important, feature. Bathymetric contours and symbols identifying rocks, wrecks, and foul ground might be more germane to plotting safe courses in and out of harbors or along the coast, but the high-water line remains an indispensable landmark for mariners and hydrographers. Although navigation systems linked to GPS satellites can track a vessel's progress across an electronic chart and warn automatically of dangerous waters, today's navigator often knows his or her latitude and longitude with greater precision than the topographer, hydrographer, and photogrammetrist who made the chart.[1] In spite of this impressive precision, or perhaps because of it, mariners have a heightened need to compare visible landmarks and cartographic coastlines. Understanding this paradox requires a short excursion into the arcane world of geodetic referencing.

Plotting locations on a chart, whether by hand or computer, requires a mathematically defined map projection that relates a symbol's position on the flat map to the feature's position on the round Earth. The projection tells the chart maker or computer where to place the reference grid of parallels and meridians, useful at sea for plotting positions known only by latitude and longitude. An appropriate projection lets the navigator read distances and

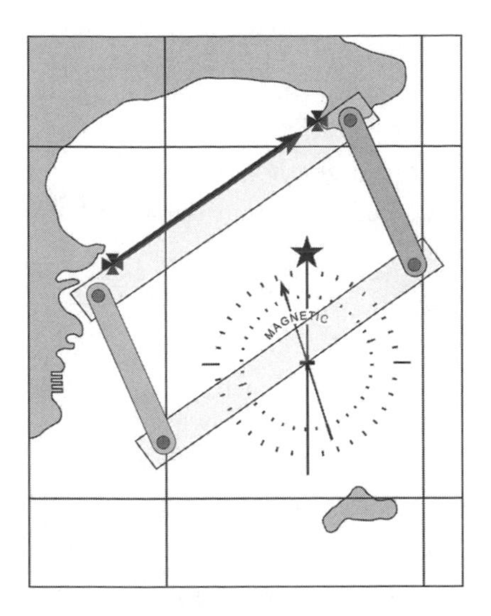

Figure 6.1. A parallel ruler is a double-hinged device that can transfer direction by keeping two straightedges parallel. One side can be aligned with a rhumb-line course (thick arrow) on a Mercator projection while the other is centered on a nearby compass rose showing bearings relative to both true north (the star) and magnetic north (reflecting the local magnetic deviation). If a compass rose is not nearby, the navigator can bridge the gap by "walking" the parallel ruler across the map.

bearings directly from the chart as well as fix a vessel's position with a sextant and station pointer (fig. 4.7), a complementary pair of instruments for measuring and reconstructing angles between prominent landmarks on the shore. A similar technique helped pre-GPS hydrographers working outward from the shoreline plot soundings on their boat sheets. By assigning each point on a chart a latitude and longitude describing its unique position on the planet, a map projection also helps chart makers compile small-scale world or regional maps from more detailed charts covering smaller portions of the coastline.

Modern charts are usually laid out on a Mercator projection, which navigators appreciate because any straight line is a rhumb line, or loxodrome, representing a course of constant direction between two points. Align a straightedge with your current and intended positions, use a parallel ruler (fig. 6.1) to transfer the direction to one of the chart's compass roses, and you have a single bearing that will take you to your destination—assuming your compass is accurate, there are no intervening hazards, and you stay on course.[2] Although mariners must cope with pesky regional, local, and transient variations in the planet's magnetic field as well as obstinate winds and currents and inconveniently placed continents, islands, reefs, and shallows, the Mercator chart works well even when a course must be broken into segments and corrected periodically. If you know where you are and can align your compass with the North Pole, the chart quickly points the way to the journey's next leg.

Introduced by Gerard Mercator as the geometric framework of his monumental 1569 world map and described mathematically by Edward Wright in 1599, the Mercator projection was largely ignored as a navigation tool until the late eighteenth century, when all other requirements for Mercator sailing were readily available: reliable compasses, the sextant, the nautical almanac, charts describing magnetic declination, and the marine chronometer, which was essential for determining longitude at sea.[3] Adopted by commercial chart makers and naval hydrographers, the Mercator grid appealed to pioneer oceanographers like Matthew Fontaine Maury, for whom mariners were both clients and informants. In adopting a Mercator framework for his Winds and Currents Charts (e.g., fig. 4.1), Maury not only promoted compatibility with contemporary sailing charts but also emulated exemplars like Robert Dudley's 1646–47 sea atlas of the known world, Edmund Halley's 1701 map of magnetic declination, and Benjamin Franklin's 1770 chart of the Gulf Stream.[4]

Despite the Mercator grid's ascendancy, the Survey of the Coast cast its charts on a different framework, the polyconic projection, conceived by Ferdinand Rudolph Hassler, its premier superintendent.[5] A surveyor and mathematician, Hassler was intrigued that a map projection could minimize distortion by dividing a region into narrow zones, elongated east to west, each with its own projection centered on a parallel of latitude. A zone's customized projection was developed on a separate cone in contact with the spherical Earth along a central parallel (fig. 6.2, *left*)—map projection works by transferring locations systematically from the globe to a flat or flattenable surface like a plane, a cone, or a cylinder. Hassler's multitude of zones, central parallels, and conic projections was advantageous because distortion increases with distance from a line of contact, called a standard parallel when the cone's apex lies along Earth's axis. Develop these mini-projections on the same globe so that their standard parallels have a common scale, and the polyconic chart will distort distance less than a chart covering the entire area on a single conic projection. Because each zone's conic projection is

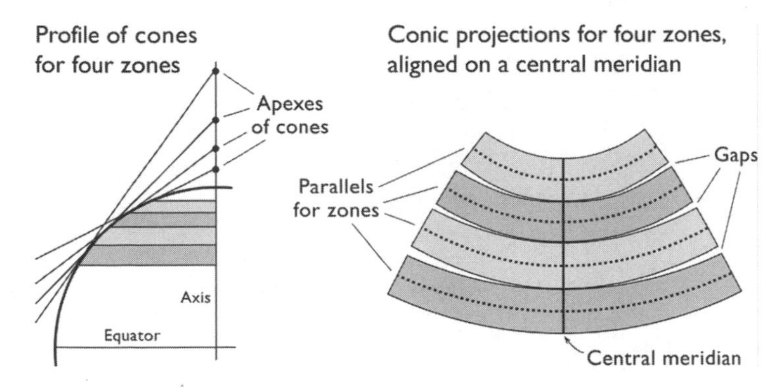

Figure 6.2. Illustration of the concept of polyconic projection for four hypothetical zones based on cones centered on the globe's axis and tangent at different latitudes (*left*) to yield a quartet of minicharts anchored to a common central meridian (*right*).

a little different, the mini-maps are aligned along an appropriate central meridian (fig. 6.2, *right*), which is also a line of true scale.

A drawback arises when a polyconic chart covers an area so large that its zones diverge noticeably along gaps that grow ever wider with distance from the central meridian (fig. 6.2, *right*). But not to worry: mathematics can conveniently suppress the gaps with an infinite number of infinitesimally thin zones. Because every parallel is now a standard parallel, east-west scale is constant, thanks to meridians that curve inward toward the poles, at opposite ends of the central meridian (fig. 6.3). Severe distortion of shape and angles along the edges no doubt explains why the polyconic projection rarely frames a world map.

Trained as a land surveyor and geodesist, Hassler was concerned more with a detailed portrait of the coastline and its accompanying hazards than with taking bearings at sea. He found the polyconic framework appealing in two ways: the projection conserved distance relationships, and with the aid of mathematical tables describing the relative spacing of meridians and parallels at various latitudes, a topographer could easily lay out a polyconic grid on his plane table as two mutually perpendicular sets of parallel straight lines. In fact, when the central meridian was placed near the local coastline, a meridian's intersection with a parallel was indistinguishable from a right angle. What's more, distances and angles could be plotted as measured: for small areas like those on a topographer's large-scale

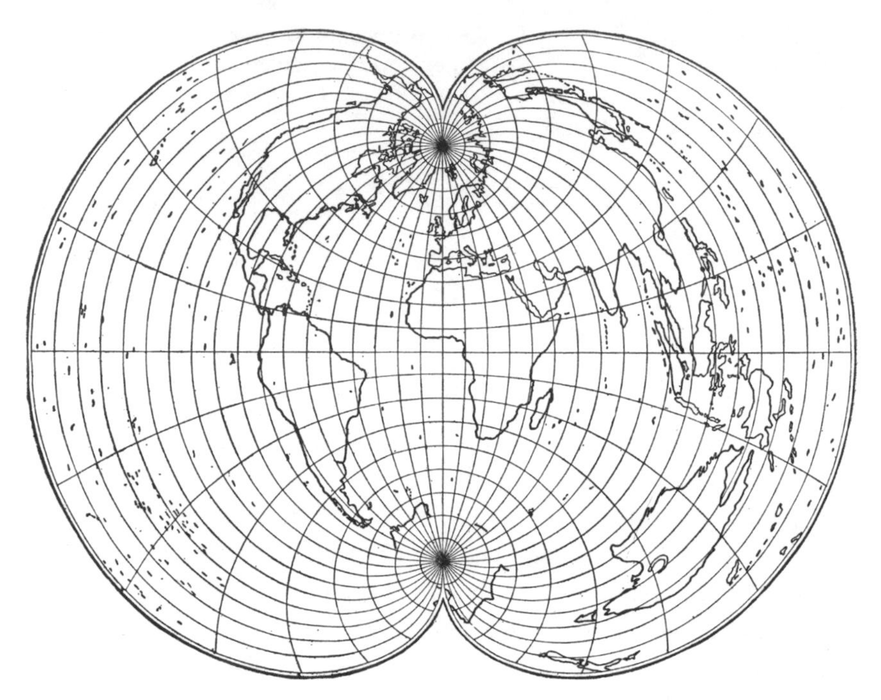

Figure 6.3. Whole-world map cast on a polyconic projection. From Deetz and Adams, *Elements of Map Projection with Applications to Map and Chart Construction*, 58.

field sheets the planet was essentially flat; for larger areas requiring curved parallels and meridians, corresponding polyconic grid lines provided a framework for transferring features to intermediate- and small-scale sailing charts. Never mind that a small-scale polyconic chart offers neither the Mercator's straight rhumb lines nor directly measurable sailing directions—a mariner could anticipate coastal hazards with one chart and plot his course with another.

Because distortions of distance and angles resulting from Earth's curvature can be virtually unnoticeable for the small areas shown on a field sheet or harbor chart, Hassler could have gotten by nicely with a Mercator grid. But Hassler was Hassler—brilliant, obsessive about precision, and uncompromisingly eccentric in dress, work habits, and geometric frameworks.[6] His choice of the polyconic grid apparently impressed his associates and successors even though he personally did little to further its use. Hassler's seminal contribution to the projection's prominence was a short discussion, less than three pages long, toward the end of a 200-page treatise published in 1825, in which he laid out his vision for the Survey of the Coast.[7] Although he touted "an assemblage of sections of surfaces of successive cones" as "the only one applicable to the coast of the United States," his proposal ignored the problem of gaping gaps between zones.[8]

Oddly enough, the Survey's first official chart, released in 1844, a year after Hassler's death, was cast on a simple rectangular projection. Equally puzzling, the agency did not publish tables and formulas for the polyconic projection until the 1850s, after which the polyconic became well established and prominently identified on American coastal charts. The Navy, which had been using the Mercator grid for decades, began to print projection tables for both grids but called attention to substantial differences between the two projections in higher latitudes. This incompatibility was not resolved until 1910, when naval officials, after years of complaint, convinced their coastal counterparts to appoint a board of experts to study the matter. The board sided with the Navy, and the Coast and Geodetic Survey (as it was then called) began recasting its charts on a Mercator framework—a slow, tedious process that took two decades.[9]

Projection tables for large-scale maps and charts must accommodate the effect of Earth's rotation which, as Isaac Newton hypothesized in 1687, flattens an otherwise spherical planet into a so-called spheroid, with a longer diameter at the equator than between the poles. This effect is akin to the centrifugal force one feels when standing on the outer edge of a carousel. With Earth turning on its axis once a day, Newton reasoned, the upward, outward force at the equator should have a small but nonetheless measurable effect on the planet's shape, so that a pole-to-pole cross-section describes an ellipse, not a circle (fig. 6.4). By the mid-eighteenth century, geodesists had confirmed Newton's hypothesis but were still trying to measure the ellipse's semi-major (longer) and semi-minor (shorter) axes. Because flattening is not severe, small-scale maps are usually based on a spherical globe, whereas large-scale maps must be developed on a spheroid, also called an ellipsoid of revolution (or more commonly, just an ellipsoid) because it is a three-dimensional figure formed by rotating an ellipse about the planet's polar axis. Although the ellipsoid requires more complicated projection formulas than a

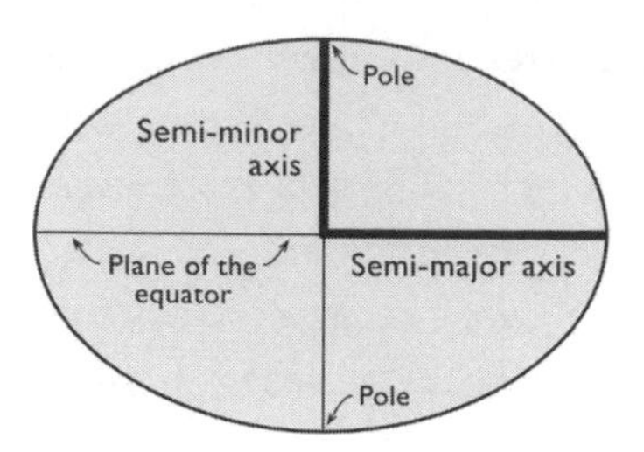

Figure 6.4. Cross section through the polar axis of an ellipsoid of revolution, described by its semi-major and semi-minor axes.

sphere, it is far more tractable than the geoid, a mathematical monstrosity that requires 32,755 coefficients to describe the undulations of the land surface and seafloor.[10]

Elongated triangulation networks covering great distances helped geodesists estimate the ellipsoid's size and shape. Called arcs of triangulation, these long, thin networks typically followed a meridian, along which Newtonian flattening is most prominent. Baseline distances, angles between triangulation stations, and estimates of latitude were mapped onto an ellipsoid, which was "adjusted" until it provided a good fit for surface measurements. The mathematical process of assigning coordinates to points described mostly by the angles and distances of a triangulation network is known as "reduction."[11] Latitude and longitude can be used to calculate distances on the curved ellipsoidal surface, and if these distances reliably match those measured in the field, the reduction is considered a good fit. An ellipsoid that received scientific approbation was adopted as a reference surface by one or more national survey organizations and used to estimate latitude and longitude for control points in more local survey networks. Once adopted, a reference ellipsoid was retained until an enterprising geodesist introduced a new ellipsoid promising a significantly closer fit. As the nineteenth century unfolded, reference ellipsoids based on longer, more carefully measured arcs typically yielded more reliable reductions than those based on older, shorter arcs in other parts of the world.

The Coast Survey based its mid-nineteenth-century projection tables on an ellipsoid described in 1841 by German mathematician-astronomer Wilhelm Bessel, who reported an equatorial radius of 6,377,397 meters (a shade longer than 3,962.73 miles in everyday measurements) and an ellipticity of 1/299.15. (Ellipticity is computed by dividing the difference between major and minor axes by the length of the former, and expressing the result as fraction with a numerator of 1.[12]) Bessel's ellipsoid was based on ten arcs of meridian, measured in Peru, India, and various parts of Europe, with a combined length of 50.5 degrees of latitude.[13] In 1880, the Coast and Geodetic Survey adopted the larger, somewhat flatter spheroid presented in 1866 by British geodesist Alexander Clarke and based on six arcs measured in Peru, India, South Africa, Russia, and western Europe, and spanning more than 76 degrees of latitude.[14] With a semi-major axis of 6,378,206 meters and an ellipticity of 1/294.98, Clarke's spheroid was bigger and flatter than Bessel's. Both ellipsoids reported a distance between the poles about thirteen miles shorter than Earth's diameter at the equator, but Clarke's estimate better fit measurements in North America, where it was widely used until the 1980s for nautical charts and the Geological Survey's topographic maps.

Replacing a reference ellipsoid with a new model is not undertaken lightly, especially after significant strides in mapping and charting would make a changeover slow and costly. American mapmakers had an opportunity in 1909, when Coast and Geodetic Survey scientist John Hayford presented an ellipsoid derived solely from North American arcs, including a coast-to-coast arc along the 39th parallel, completed in 1898. Although Hayford's ellipsoid offered a marginally better fit, it was passed over in 1913, when the United States, Canada, and Mexico agreed to use the Clarke ellipsoid for a continent-wide readjustment of triangulation measurements, later named the North American Datum of 1927.[15]

As the need for a separate technical term suggests, a datum—or more precisely, a *horizontal* datum, to distinguish it from a vertical datum like mean lower-low water, discussed in chapter 2—is more than just a reference ellipsoid. To provide the best possible fit for triangulation measurements on a particular continent, some datums offset the center of the ellipsoid from the center of the planet. Figure 6.5, which describes this offset schematically, highlights another geodetic paradox, the European Datum of 1950, which is based on Hayford's ellipsoid, also known as the International Ellipsoid, even though it was derived solely from arcs crisscrossing North

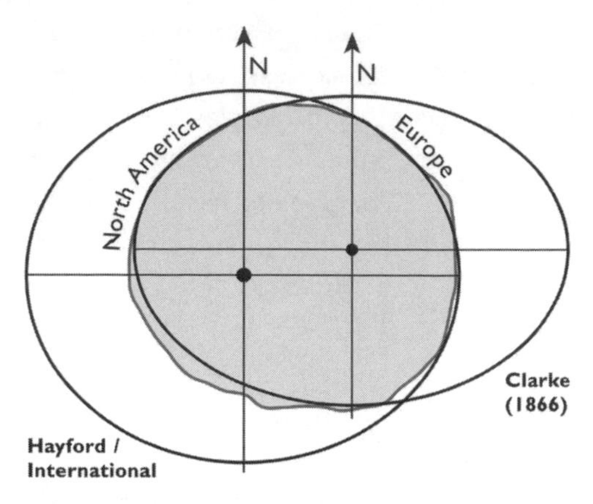

Figure 6.5. Schematic diagram emphasizing differences between reference ellipsoids for the North American Datum of 1927 and the European Datum of 1950. Each ellipsoid is offset differently from Earth's center to provide a regionally optimal fit for angles and distances within the region. Short axes of both ellipsoids are parallel to Earth's axis. The diagram greatly exaggerates bulges in the geoid (shaded, irregular figure), which is not symmetrical about the planetary axis. Ellipses are grossly exaggerated to underscore the facts that the Clarke Ellipsoid of 1866 is flatter and the Hayford/International Ellipsoid is larger.

America.[16] Suitably offset, Hayford's and Clarke's ellipsoids provide good regional approximations of the more complex geoid, which bulges differently in Europe and North America.

A datum not only affects the latitudes and longitudes assigned to survey monuments and other markers used for large-scale mapping, it also determines the precise positions of meridians and parallels, including those that frame topographic quadrangle maps. As users of USGS topographic maps discovered in the 1980s, a new datum can shift sheet boundaries noticeably. When the North American Datum of 1983 (called NAD 83 for short) replaced NAD 27, sheets initiated before final calculations were framed by the old datum and showed only the "predicted" locations of the quadrangle's new corners, typically marked by intersecting dashed lines, as illustrated by an excerpt from the Mercer Lake, Oregon map (fig. 6.6).[17] As fine print at the bottom of the sheet explains, "To place on the predicted North American Datum of 1983, move the projection lines as shown by dashed corner ticks (23 meters north/98 meters east)"—a net displacement of 101 meters (330 feet), which underscores the danger

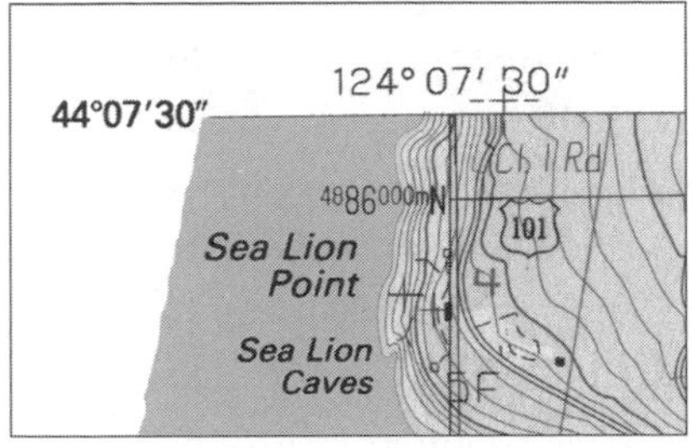

Figure 6.6. Excerpts from upper-left corner of the U.S. Geological Survey's Mercer Lake, Oregon, 7.5-minute topographic map, provisional edition, published in 1984 at 1:24,000, with the portion at the top enlarged to highlight the "predicted" location (dashed lines) of the quadrangle corner on NAD 83. Map extends outside the quadrangle to accommodate a small portion of the coast beyond the sheet's normal boundary at 124° 07′ 30″ W.

of using latitude and longitude to transfer positions between charts based on different datums.

Although nautical charts were affected by adoption of NAD 83, the shift was less problematic for mariners wary of mixing old and new charts.[18] Frugal boat enthusiasts using older charts based on NAD 27 could get by nicely as long as they stuck with similarly obsolete charts and checked the Coast Guard's Local Notice to Mariners, a monthly (now weekly) announcement of restrictions, hazards, and changes to channel markers and other navigations aids.[19] More worrying was satellite positioning, a Defense Department project that offered a continuously updated fix based on a constellation of twenty-four Navstar satellites.[20] GPS signals reference location to the World Geodetic System 1984 (WGS 84), an intercontinental datum designed for self-guided weapons like the Cruise Missile.[21] Tied to the planet's center of mass, rather than offset like NAD 27 and ED 50, WGS 84 yields latitudes and longitudes virtually indistinguishable within North America from those registered to NAD 83, which is also geocentric. But as the datum shift in figure 6.6 implies, it's risky to plot WGS 84 positions on NAD 27 charts.

Mariners appreciate GPS for its remarkable accuracy in monitoring a vessel's progress across an electronic chart showing coastlines, channel markers, navigation hazards, and other features of the paper chart. Although satellite positioning must cope with atmospheric interference and various electronic limitations, it's a lot more reliable than older positioning techniques that rely on stars or radio beacons.[22] As originally introduced in the late 1980s, satellite tracking promised military users positional estimates accurate to within 21 meters (69 feet) 95 percent of the time.[23] Differential GPS (DGPS) reduced the probable error to 2 to 7 meters by applying a real-time correction broadcast from a receiver with a precisely known location—the instantaneous adjustment needed to correct discrepancies at the DGPS station works well for other GPS receivers within about 300 miles.[24] A deliberate degradation called Selective Availability (SA) widened the 95-percent radius of uncertainty to 100 meters (330 feet) for civilian users, but in May 2000, the Defense Department turned off SA after persuasive lobbying by the consumer electronics industry, eager to sell GPS receivers to hikers, motorists, and recreational boaters. With a DGPS signal provided by the Coast Guard or a correction called Wide Area Augmentation Service (WAAS) provided by the Federal Aviation Administration, mariners in coastal waters typically enjoy military-level accuracy of 3 meters (10 feet) or better.[25]

A likely error of 3 meters is especially impressive when compared to the positional uncertainty of channels, navigation markers, coastlines, and various hazards shown on nautical charts. Rooted in the challenges of measuring angles and distances, positional errors on nautical charts also reflect the more mundane reality of graphic symbols appreciably thicker than the point and line features they represent. There's little point, nautical cartographers reasoned intuitively, in greater precision than a chart can show—better to spend the money on expanding coverage and keeping charts up to date. But as measurement techniques improved, chart makers devised accuracy "standards" based on map scale and line width. When federal chart makers started thinking about these matters, 1.5 millimeters ($\frac{1}{17}$ inch) became the "specified chart accuracy," defined as "the accuracy with which features are plotted on the chart from their original surveyed position."[26] As a standard for uncertainty, 1.5 millimeters equates to 30 meters at 1:20,000 and 120 meters at 1:80,000, which is a concise way of saying that at 1:80,000 a line 1.5

millimeters wide takes up as much space as a precisely portrayed corridor 120 meters (400 feet) wide, even though the line itself might represent a much narrower feature like the high-water shoreline. Uncertainty arises because the shoreline could occur anywhere within this corridor. In the mid-1990s, when differential GPS allowed more precise positioning of field data, NOAA tightened the standard to 1 millimeter ($\frac{1}{26}$ inch), which reduced uncertainty to 20 meters at 1:20,000 and 80 meters at 1:80,000.[27] Additional imprecision arises from "cartographic license," which lets chart makers shift symbols slightly for clarity.

Fully understandable as a limitation of paper maps, this graphic ambiguity is embedded in electronic navigation charts designed for use with GPS. In the mid-1990s NOAA scanned its entire set of a thousand paper charts and made their images available on CD-ROMs as Raster Nautical Charts (RNCs) compatible with electronic systems displaying the portion of the chart surrounding a you-are-here icon representing the boat's position.[28] GPS tracks the vessel's movement, while the chart display system continuously re-centers the image, stored electronically like a giant digital photo, with thousands of rows, or rasters, of pixels (picture elements). Scanning was a joint venture with Maptech Inc., a firm specializing in geospatial data and GPS-integrated viewing devices. A CD-ROM contains images for about fifty charts and can be used with display systems from several companies. According to NOAA, raster charts outsell their paper counterpart eight to five.[29] Mariners preferring paper charts can buy them from marine supply stores and other vendors, order them by mail from the Federal Aviation Administration (which also distributes aeronautical charts), or purchase a print-on-demand (POD) version from OceanGrafix, a commercial partner with a network of dealers equipped with high-resolution, large-format printers.[30] POD charts offer the look and feel of a conventional paper chart and incorporate the latest corrections. Mariners eager for the most current information update their raster charts with a weekly e-mail containing links to small image files called "update patches," used by the chart display system to overwrite obsolete portions of the raster image.

As a navigation tool, the chart display system seems a decided improvement over a cumbersome 48-by-37-inch paper chart. A mariner can follow the chart hands-free or plan a course by zooming out, panning forward, and zooming in for a closer look at symbols representing rocks, wrecks, or narrow channels—there's little risk as long as the user obeys channel markers and doesn't venture too close to questionably positioned hazards. Zooming in too closely might be misleading because the chart image, originally scanned at 762 ppi (pixels per inch) but distributed at a lower resolution (254 ppi) to fit more charts onto the CD, allows a bit of "overzooming" (perhaps as coarse as 72 ppi) before the image on display screen starts to look blocky.[31] Mariners are well advised to heed navigation guru Nigel Calder's warning that "the user of any chart should not be lured into a false sense of security about its accuracy."[32]

Misplaced faith in the chart maker's authority can be demonstrably riskier with the Electronic Navigation Chart (ENC), a NOAA database designed for use with an Electronic Chart Display and Information System (ECDIS), which can monitor a vessel's proximity to reefs, rocks, and channel boundaries as well as display their shape and position on a screen.[33] Unlike the fine-grained row-and-column structure of a raster chart's pixels, an ENC consists of vector data, the cartographer's term for lists of point coordinates describing linear features like coastlines, channel boundaries, and bathymetric contours. A typical vector list treats a region of shallow water or a restricted area like a gunnery range as a polygon bounded by straight-line segments linked together at the points in the list. In contrast to the raster chart's static image, which demands visual inspection and informed interpretation, a vector chart enables real-time monitoring because software can vigilantly compare hazard polygons with the vessel's position, constantly updated by the GPS—cross a boundary into a danger zone, and the ECDIS triggers an audible warning. And because software can generate buffers around hazard zones and point out features like rocks and wrecks, an ECDIS can warn mariners away from known dangers described by the database. Buffers are helpful because some of the data are less reliable than the GPS.

A close look at paper charts and their electronic equivalents reveals striking similarities in their coastlines and other delineations. Figure 6.7 illustrates these similarities for the coastline near Five

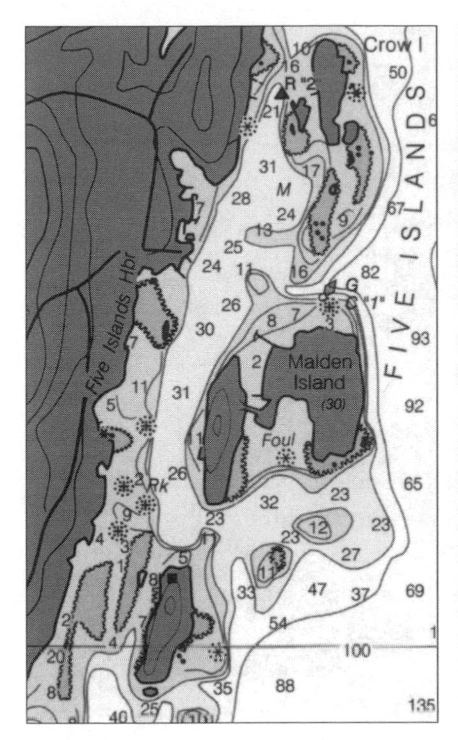

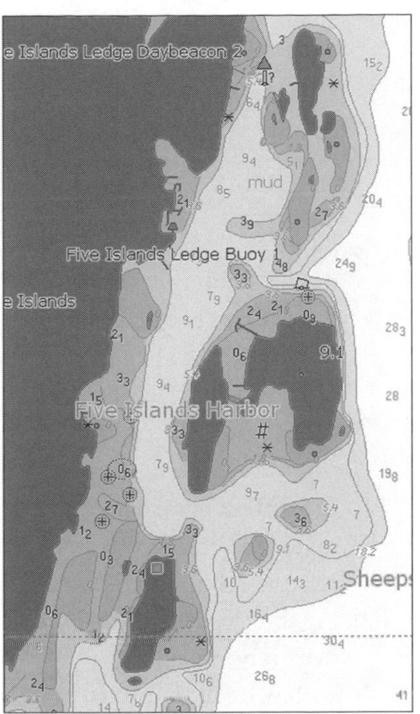

Figure 6.7. Juxtaposition of paper (*left*) and ENC (*right*) representations of the coastline at Five Islands, Maine. Both images are excerpts of NOAA chart 13295, Kennebec and Sheepscot River Entrances. Scanned image of the 1:15,000 paper chart, 11th edition (October 2002), is similar in content to a Raster Navigation Chart (RNC).

Islands, Maine. The left-hand excerpt, scanned from a paper chart, also reflects coastal delineations on the corresponding raster chart, scanned at a somewhat lower resolution from a hardcopy image. By contrast, the example to its right shows the detail displayed electronically by "chart viewer" software that generates a visible image from the vector data of an ENC.[34] Enhanced in PhotoShop to restore some of the contrast lost in converting from color to black-and-white, these excerpts show markedly similar land-water boundaries—hardly surprising because the coastline on the ENC was traced from a similarly scanned image. The principal differences on the vector-based image are soundings in meters rather than in feet, simple lines instead of the traditional scalloped symbol marking the perimeter of rocky areas covered at high tide, an absence of roads and elevation contours above the shoreline, and consistently horizontal labels outlined in white to enhance legibility against a darker background.

Unlike raster and print-on-demand charts, which are sold by commercial partners, ENCs are distributed free over the Internet, a policy that underscores NOAA's new long-term goal of producing a single, vector-based navigation-safety product and leaving the paper-chart business to the private sector. However impressive, today's vector charts are very much an interim solution, integrating less precise but nonetheless useful renderings of coastlines scanned from traditional charts with more geodetically precise descriptions of navigation channels, obstructions, and aids to navigation.[35] Although an ENC can be less noticeably and thus more dangerously magnified than a raster chart, smart-chart display systems promise an explicit communication of uncertainty, through inset or pop-up maps describing the sources for various parts of a chart and their relative reliability as well as "zone of confidence" symbols advertising the positional uncertainty of specific features.[36] Because mariners rely on charts for selecting a course as well as navigating, buffers need to be visual as well as numerical.

Knowing what you're getting into is particularly important for rocky shores like Five Islands, where navigation on autopilot could easily become a nightmare. As a close look at the twin charts in figure 6.7 suggests, mariners unfamiliar with local waters must move cautiously, in clear weather as well as in fog, and rely on sonar and radar, if they have them, to avoid submerged hazards and other vessels. Although charts can warn of some dangers, cautious mariners also consult tide tables, the Coast Guard's Light List, and NOAA's Coast Pilot, all three included on the raster-chart CD and available free on the Internet in their most up-to-date versions.[37] The Light List reports the locations and characteristics of buoys, channel markers and other guideposts, while the Coast Pilot supplies geographic details too specific for the nautical chart's standardized graphic vocabulary. An electronic "smart chart" might expand upon its symbols in an optional pop-up or on adjacent screen, but no paper chart, RNC, or ENC can communicate quite as concisely what three paragraphs in the Coast Pilot reveal about Five Islands:

Five Islands Harbor, a narrow passage between Five Islands and the western shore north of Dry Point, forms a secure harbor for small craft, with depths of 18 to 30 feet. The main entrance is northward of Malden Island, the largest wooded island, which is 30 feet high. A colony of summer homes is on the island, and a private float landing is on its northwestern side. **Malden Island** is connected to the island close westward of it by a bridge. In the middle of the entrance is a rock covered 11 feet and marked by a buoy. In entering, craft can pass the buoy close-to on either side, but the best water is reported to be on the north side.

Boats also can enter the harbor from the northwestward, following the western shore and passing inside of all islands and shoals. **Crow Island Ledge,** extending west from Crow Island at the northern entrance, is marked by a daybeacon. Northwestward of the daybeacon, an unmarked ledge makes out from the Georgetown Island shore. Care should be taken to avoid it by favoring the Crow Island side of the channel slightly and passing close westward of the daybeacon. The southern entrance, nearly blocked by rocks and ledges that uncover about 4 feet, should not be used without local knowledge. There is also a clear channel from the eastward south of Malden Island.

Five Islands is a village on Georgetown Island on the western side of the harbor. There are several float landings. A marina has depths of 6 to 10 feet reported alongside its float landings. Transient berths, gasoline, and some marine supplies are available. A 10-ton fixed lift can handle craft up to 40 feet for hull and engine repairs or dry open or covered winter storage. The village landing, adjacent southward, has 12 feet alongside. Provisions can be obtained at a store at the landings, and there is a snack bar.[38]

As this excerpt demonstrates, a coastline embodies conveniences as well as hazards.

SEVEN

GLOBAL SHORELINES

Ever wonder about the connection between the coastlines in your world atlas and the comparatively detailed shorelines on large-scale navigation charts? While there's surely a link, don't expect a common ancestor for page-size world maps. Cartographers customarily compile smaller-scale maps from larger-scale sources—that's the most reliable way to make less detailed maps of wider regions—but what they use depends upon what's available, which can vary widely because of different practices and priorities. Although geographers and mapmakers have recognized the value of international cartographic cooperation for over a century, efforts to get all the world's countries on the same projection, if not the same page, illustrate the political, economic, and technical difficulties of producing a coherent set of global base maps.[1]

The first serious attempt at a moderately detailed global map with standardized symbols and content is the International Map of the World (IMW) proposed in 1891 by German geographer Albrecht Penck at the Fifth International Geographical Congress, held at Berne, Switzerland.[2] Echoing the optimistic internationalism of late-nineteenth-century science,[3] Penck called for a worldwide series of map sheets published at a scale of 1:1,000,000 and uniform in projection, symbols, style, and technique. Compiled from existing or soon-to-completed maps at varied scales, the millionth map (as it came to be known) would adequately describe the world's mountains, rivers, and key coastal features without incurring prohibitive production costs. Comparatively detailed 1:200,000 mapping already covered most of Europe, much of North America, and significant portions of Africa and India, he noted in an 1893 essay in the *Geographical Journal,* and "more than two-thirds of the whole extent of coast-line of our continents and islands are mapped on a not much smaller scale."[4]

Intrigued with the idea, the Berne delegates appointed an international commission to study the proposal, which Penck enthusiastically expanded.[5] In addition to standardizing symbols, colors, and elevation categories (separated by contours at 100, 300, 500, and 1200 meters), he divided the world into 880 sections encompassing five degrees of latitude and five degrees of longitude, based on the Greenwich Meridian, adopted as the international standard in 1884 but resisted by the French until 1911. The number of map sheets would have been larger had Penck not assigned isolated islands to other sheets as inset maps and extended 49 sections slightly east or west and another 27 slightly north or south. Poleward of 60 degrees latitude, "double-column" sections covering 10 degrees of longitude further reduced the number of maps sheets to 769. Penck assigned sections to specific countries based on homeland territory, overseas possessions, and history of scientific exploration. Over half the sheets went to the British Empire (222) and Russia (192), with the United States (65) and France (55) ranking third and fourth. At the bottom of the list were Belgium, Switzerland, and Greece, with one sheet each. In estimating production cost, Penck calculated that the finished map, if laid out as a mosaic, would cover 2,127 square feet—about half the size of a basketball court. Although an expected deficit greater than £100,000 was clearly "a large sum," the project was no more costly than other massive scientific undertakings of the day, such as meteorological measurements in the Arctic or the detailed map of the heavens proposed by the International Astronomical Congress.[6]

Participating nations were supportive, but chauvinistic nitpicking delayed development. When the International Geographical Congress (IGC) met again, in London in 1895, the steering committee discussed numerous suggestions, including predictable proposals from Britain, which preferred heights and distances measured in feet and miles, and France, which held out for longitudes anchored at Paris, not Greenwich.[7] Although the millionth map was on the agenda of subsequent IGC meetings in Berlin (1899), Washington, D.C. (1904), and Geneva (1908), not much happened until November 1909, when national mapping officials invited to London by the British Government overhauled and expanded Penck's original plan.[8] Single-column map sheets would cover 4 degrees of latitude and 6 degrees of longitude, a modified polyconic projection guaranteed a perfect match along the edges of adjoining sheets, and mean sea level "deduced in each country from tidal observations on its own records" provided a workably flexible datum for elevation contours and isobaths.[9]

Britain, France, and Hungary issued the first official millionth-map sheets in 1911, and the United States, which sent two Geological Survey staffers to the London conference, published its first sheet the following year. Covering about eight times as much water as land and named for its most prominent city, the Boston sheet included parts of all New England states except Vermont. The excerpt in figure 7.1, centered near Five Islands, Maine, illustrates the contribution of the U.S. Coast and Geodetic Survey, which provided the map's shoreline as well as soundings for its bathymetric contours. William Joerg, who reviewed the sheet in the *Bulletin of the American Geographical Society,* praised its 100-meter isobaths for providing "a clearer picture of the continental shelf in this region than any other heretofore available."[10] That America's coastal cartography was more complete than its interior topography was underscored by the compilers' reliance on 1:158,400-scale atlas maps as the key source for southeastern New Hampshire, largely neglected by USGS topographers. That you can't easily make out Maine's Sheepscot River, much less the harbor at Five Islands, highlights the extreme generalization of coastal details at a scale of 1:1,000,000.

As the crowded delineations in figure 7.1 suggest, Geological Survey mapmakers were reluctant to sacrifice hard-won details, especially for a new map series with uncertain sales. In an article published two years earlier in *National Geographic Magazine,* Bailey Willis, one of two USGS officials at the London conference, reported that

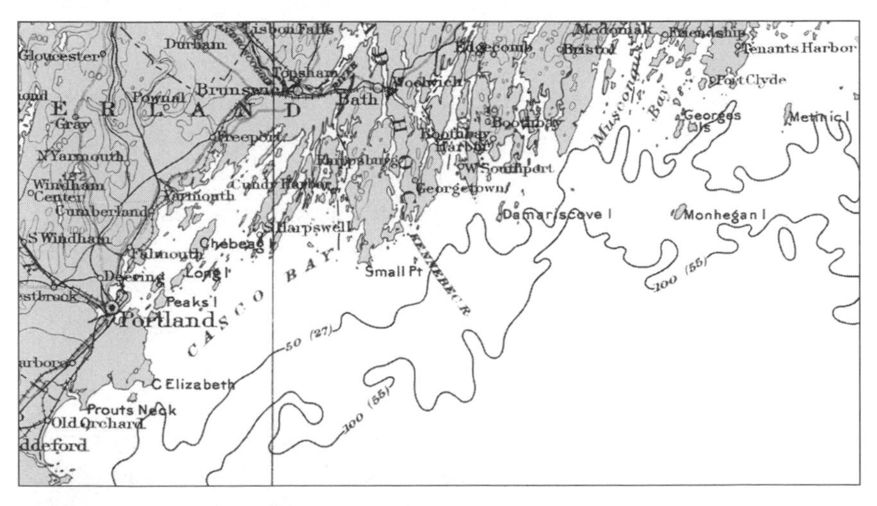

Figure 7.1. Portion of U.S. Geological Survey map sheet North K-19, Boston, published in February 1912 as the first United States contribution to the International Map of the World at a Scale of 1:1,000,000. This example, taken from the top of the map, is centered longitudinally around Five Islands, Maine.

compilation was underway on the first nine American sheets—sort of. Because of the vast difference in detail between the millionth map and existing 1:62,500 and 1:125,000 source materials, on which an inch represents one and two miles, respectively, USGS compilers created intermediate 1:500,000-scale maps, drawn at "8 miles to the inch . . . in such a manner that they may be reproduced by photolithography in a clear and effective manner for publication on a scale of 10 miles to the inch."[11] Compiling at an upsize was a common cartographic practice—it was much easier to correct mistakes as well as position symbols precisely—but it also allowed the Geological Survey to create another product for a presumably eager domestic audience. As Willis revealed, "in this form the maps may become immediately available for use by the departments of the government or by individual states." That the millionth map was a lower priority is apparent in his observation that "eventually, as Congress provides the means, they will be engraved and published on the scale of one million (16 miles to the inch)."

For whatever reason—a stingy Congress, more pressing needs, or the creeping isolationism that would derail the United States' participation in the League of Nations—interest in the International Map waned rapidly. Although Willis reported nine map sheets as "in preparation,"[12] the Geological Survey issued only two additional maps that decade: the San Francisco Bay sheet in 1914 and the Point Conception sheet (also in California) in 1915.[13] The next American installment, the Hudson River sheet, did not appear until 1927, three years after President Calvin Coolidge endorsed a formal request from the secretary of state that Congress appropriate $30 as "a contribution by the United States toward the secretarial expenses" of the international coordinating bureau.[14] At least the country was starting to pay its dues.

That the first four maps were coastal sections probably reflects an eagerness to get the easiest sheets into print first rather than an appreciation of the Coast and Geodetic Survey's contribution. In 1931 Arthur Hinks, a prominent British geographer with a strong interest in the millionth map, denigrated the American contribution as "four in all and mostly at sea."[15] The four maps, he figured, were the topographic equivalent of only one and a half sheets.[16]

Supporters of the International Map of the World faced two challenges: getting national mapping agencies to participate and implementing the international standards refined in Paris in December 1913 by representatives from thirty-four countries. Recognizing the need for coordination and oversight, the Paris delegates authorized a Central Bureau, maintained by Britain's Ordnance Survey (OS), to coordinate development and prepare annual progress reports. World War I thwarted a subsequent meeting, planned for Berlin the following year, and disrupted compilation more generally, except for the harried, expedient cartography inspired by urgent wartime demands for geographic intelligence. Between 1914 and 1918 the Royal Geographical Society, with Hinks's guidance and military support, produced over a hundred map sheets for the "Europe 1/Million Provisional Series," which covered much of Europe and parts of North Africa and the Middle East, and conformed to the scale and section boundaries of the millionth map.[17] The "provisional" label reflects admitted deficiencies in content and aesthetics.

Although the provisional map was useful for military planners and postwar diplomats, upgrading the sheets was not Britain's responsibility. A 1921 survey by the Central Bureau, up and running at OS headquarters in Southampton, counted only twenty-nine map sheets in print worldwide, with perhaps another fifty "in preparation." [18] Four years later, the Bureau estimated that two hundred of the roughly nine hundred non-oceanic sections had been mapped "in some form or other" at the one-millionth scale but "only twenty-one of these conform[ed] exactly to the Paris resolutions." [19]

One-to-a-million mapping apparently made a strong impression on Isaiah Bowman, who attended the 1919 Paris Peace Conference as head of a group of American geographers and mapmakers known as "The Inquiry." Headquartered in New York at the American Geographical Society (AGS), which Bowman directed, the group was formed with the support of President Woodrow Wilson, who wanted a strong United States presence at the negotiations.[20] In late 1918 an Inquiry staff of 150 assembled a huge set of maps and related materials, which Bowman and two dozen others escorted to Paris, where they helped reconfigure the political map of Europe and the Middle East. In March 1920, at Bowman's suggestion, the AGS launched an ambitious project to map Latin America "on the scale of 1:1,000,000 which will conform to the scheme of the International Map." [21] Bowman, who was a Latin Americanist, claimed to have conceived the idea for a compiled map of Latin America "in early 1913 or late in 1912," at a meeting with AGS staff;[22] appointed director in 1915, he was unable to follow up until after the war. Persuasive as both scholar and fund-raiser, Bowman convinced the AGS Council that mapping at the one-millionth scale was not only useful and efficient but a significant contribution to an international project largely ignored by Central and South American governments.

Completing the 107 map sheets took twenty-five years and over a $500,000, mostly from a handful of wealthy AGS Council members.[23] Wary of usurping the role of official mapmakers in Latin nations, the AGS added the label "Provisional Edition" to its map sheets and steadfastly pointed out that the Map of Hispanic America on the Scale of 1:1,000,000 was not part of the International Map. Even so, countries south of the Rio Grande cooperated by providing unpublished surveys, and a few even used the AGS map as a framework for settling boundary disputes. Although not a part of the official

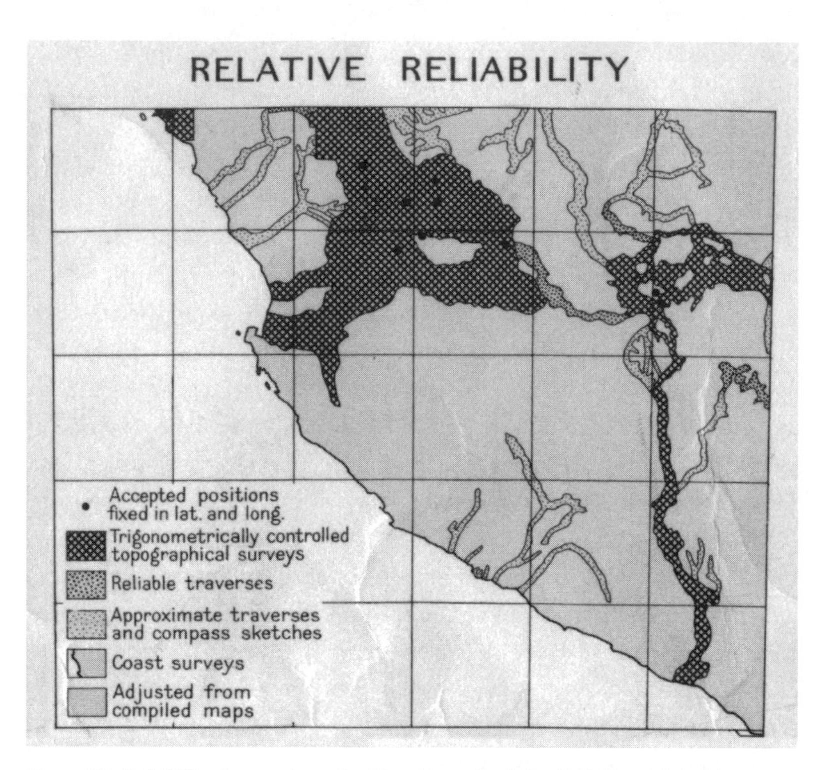

Figure 7.2. Reliability diagram from the Lima, Peru sheet, South D-18, published in 1938 by the American Geographical Society as part of its Map of Hispanic America on the Scale of 1:1,000,000. Excerpt was enlarged 50 percent (from 1:10,000,000 to 1:6,666,666) for clarity.

millionth map, the Map of Hispanic America followed official guidelines and contributed an important improvement: the Relative Reliability diagram (fig. 7.2) added in the margin when the quality of source materials varied markedly across a map sheet.[24] Completed in 1945, the project impressed the Royal Geographical Society, which three years later honored Bowman for (among other achievements) initiating "the greatest map ever produced of any one area" and "a monumental feat in furtherance of geography."[25]

Although accurate source materials were comparatively abundant north of the Rio Grande, the U.S. Geological Survey issued no new installments between 1927, when it published the Hudson River map sheet, and 1942, when the Chesapeake Bay sheet appeared. After publishing nine more maps between 1948 and 1952,

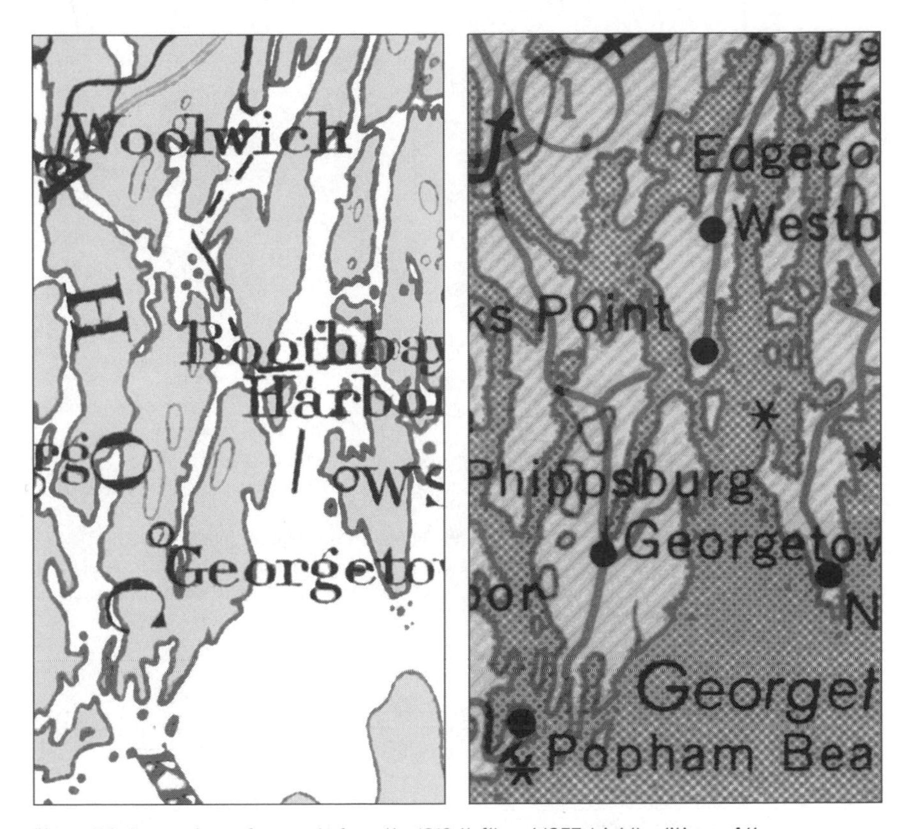

Figure 7.3. Comparison of excerpts from the 1912 (*left*) and 1957 (*right*) editions of the Boston millionth-map sheet, enlarged from 1:1,000,000 to 1:250,000. Area shown is centered horizontally on Five Islands, Maine, located just above the rightmost prong of the *u* in Phippsburg on the right-hand excerpt.

the USGS ceded responsibility to the Army Map Service, which was developing a nationwide topographic map at 1:250,000 and had the personnel and funding to expedite one-millionth-scale coverage of the conterminous forty-eight states.[26] Although military mapmakers finished the job, the Geological Survey disparaged their sheets as "provisional editions" that "do not conform to all IMW specifications."[27] One of the Army's first contributions to the IMW was a revision of the Boston sheet, published in 1957. As enlarged excerpts for the vicinity of the Sheepscot River illustrate (fig. 7.3), the 1912 and 1957 editions reflect different, but not radically dissimilar, generalizations of Maine's saw-tooth coastline.

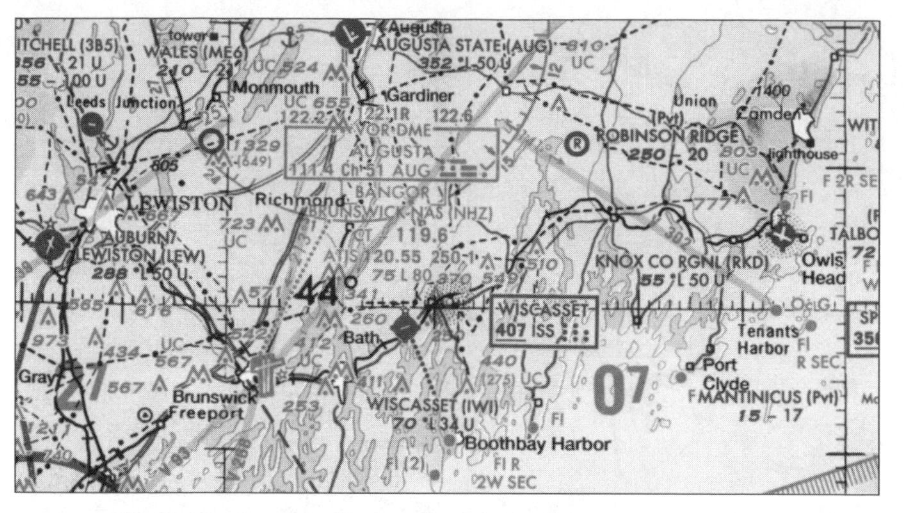

Figure 7.4. Portion of 1:1,000,000-scale World Aeronautical Chart sheet CF-19, 35th edition, April 14, 2005, published by the U.S. Department of Transportation, Federal Aviation Administration, National Aeronautical Charting Office. Five Islands, Maine is near the *A* in WISCASSET, in the bottom center of the map.

A more noteworthy transition occurred in late 1953, when the Cartographic Section of the newly formed United Nations took over the coordinating role of the Central Bureau.[28] Eager to re-energize a struggling endeavor less than half complete, the UN hoped to fill in blanks spaces on the IMW status map while resisting comparisons with analogous endeavors like the 1:2,000,000 map of Africa initiated by the Royal Geographical Society and the 1:1,000,000-scale World Aeronautical Chart (WAC), promoted by the International Civil Aviation Organization and far more complete. As the WAC excerpt in figure 7.4 shows, base map features relevant to visual navigation and flight planning are visually subordinate to symbols describing airport runways, obstructions, radio aids to navigation, and designated airspaces. Designed to meet an important, well-defined need, the WAC was essentially complete by 1945.[29]

Doubts about the need for two separate international map series intensified in 1962, when a conference convened by the UN to update standards for the official millionth map recommended replacing the IMW's modified polyconic projection with the WAC's Lambert conformal conic projection, which minimizes distortions of great-circle flying routes. Arthur Robinson, a prominent American cartographer, intensified these misgivings two years later, at a meeting of the International Geographical Congress, when he asked, "What other purposes can be served by the 1:1,000,000 International Map beyond that of providing reconnaissance coverage and cartographic wallpaper?"[30] Despite its agreeably rounded denominator, 1:1,000,000 was "too small for one to plot field observations" and "a very difficult scale at which to generalize various kinds of boundaries." A better solution, he argued, was a 1:2,000,000-scale map "specifically designed for a base map" and conscientiously "kept up to date."[31] Undeterred, the UN coordinators soldiered on, adding a few more sheets every year to their IMW status map. In 1987, with Albrecht Penck's millionth map looking more like a white elephant than a sacred cow, the UN accepted a study committee's recommendation that it no longer support the series.[32]

Mapping the world's coastline at a uniform scale with standardized symbols proved easier when the focus was water, not land. The first and most significant international hydrographic collaboration was the General Bathymetric Chart of the Oceans, better known by its acronym GEBCO.[33] Initiated in 1903 by Prince Albert I of Monaco, whose avocation was marine science, GEBCO proved more practicable than the International Map of the World because its territory was truly international and its significantly smaller scale, 1:10,000,000, allowed global coverage with far fewer map sheets. The project was well within the capability of a single institution, which was formally established in Monaco in 1921 as the International Hydrographic Organization.[34]

GEBCO's first edition, published in 1905 as *Carte générale bathymétrique des océans,* consisted of a title sheet (in French), twenty-four map sheets, and an index sheet describing their relative positions (fig. 7.5).[35] Centered on the north and south poles, a pair of gnomonic projections, on which straight lines represent great circles, framed the eight sheets covering the Arctic and Antarctic realms. Between 72°N and 72°S a Mercator projection, favored by navigators, framed the other sixteen sheets. Because of the Mercator map's notorious but purposeful distortion, scale varied from 1:10,000,000 at the equator to 1:3,100,000 at 72°.[36] The most prominent line within the

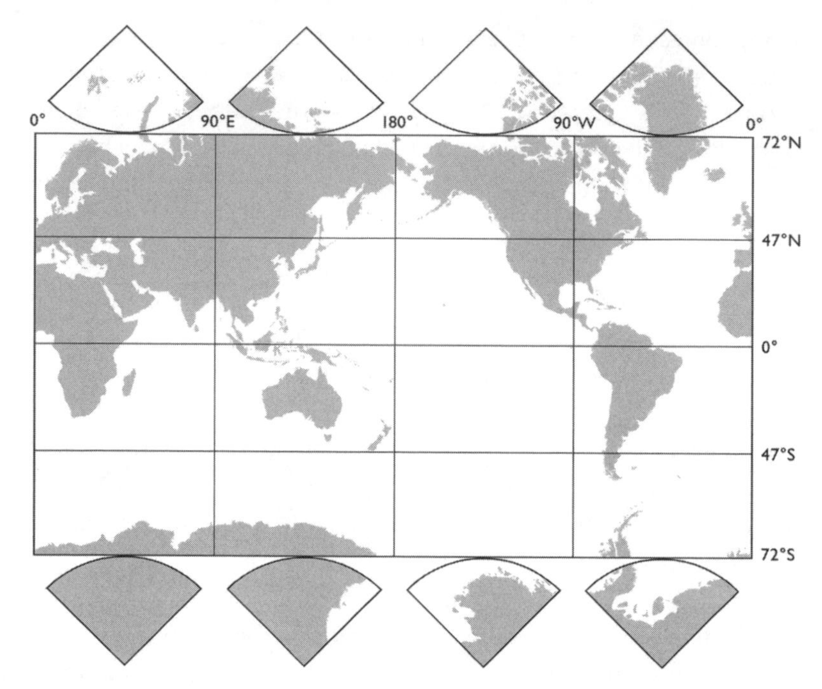

Figure 7.5. Layout diagram for the first edition of the General Bathymetric Chart of the Oceans, issued as a set twenty-four map sheets in 1905. Adapted from figure 9 in Scott and others, *The History of GEBCO 1903–2003*, 24.

hastily produced map was its coastline, further highlighted by the buff-colored tint of its largely featureless land areas—a minor complaint of critics irritated mostly by erroneous soundings and inaccurate feature names. Embarrassed but not discouraged, Prince Albert commissioned a much improved second edition, which took two decades to complete.[37] Third and fourth editions initiated in 1929 and 1954 reflect further increases in bathymetric detail and named undersea features.[38]

Prince Albert's cartographers were understandably vague about tidal datums. Although their source materials reflected diverse conceptions of the shoreline, positional discrepancies in the chart's land-water boundary were inconsequential. At 1:10,000,000, for instance, a thin, $\frac{1}{50}$-inch-wide line is the equivalent of a corridor more than three miles across on the ground—ample wiggle room to mask horizontal differences between mean high water and mean lower-low water, which are less problematic at this scale than the displacement and smoothing needed to make a chart's coastline and other symbols look good.

Committed to undersea exploration and international collaboration, the IHO joined the United Nations Intergovernmental Oceanographic Commission (IOC) in producing a fifth edition, published in 1982 with accompanying text in English as well as French.[39] Electronic tracings of its 1:10,000,000 coastlines and isobaths served as a foundation for the GEBCO Digital Atlas, released in 1994 and reissued in 2003 as the *Centenary Edition of the IOC/IHO General Bathymetric Chart of the Oceans*.[40] CD-ROM publication allowed the addition of several related datasets, including the World Vector Shoreline (WVS), developed in the 1980s by the U.S. Defense Mapping Agency.[41] Intended for display at scales as large as 1:250,000, the WVS database is registered to the World Geodetic System 1984 datum (p. 79), which also provides the geographical framework for the Defense Department's GPS satellites.

As the "vector" in its name implies, the World Vector Shoreline treats the coastline as a sequence of point coordinates (latitude-longitude pairs), similar to the vector structure of the Electronic Navigation Chart, examined in the previous chapter. While coastal delineations in both databases can be traced back to paper charts, the WVS was derived directly from the Defense Department's Digital Landmass Blanking Data, developed to help radar reconnaissance aircraft operating along the coast improve their detection of ships and planes by filtering out spurious reflections from terrain.[42] The landmass data treat the planet's surface as a matrix of numbers (0 for water, 200 for land) recorded at intersection points in a network of meridians and parallels spaced 3 arc-seconds apart (fig. 7.6). (An arc-second is $\frac{1}{3,600}$th of a degree, that is, $\frac{1}{60}$th of a minute, which is $\frac{1}{60}$th of a degree.) Sampling points are spaced about 300 feet (92 meters) apart along the meridians, but are generally closer along the parallels, which shrink in circumference as latitude increases. To create the World Vector Shoreline, a computer "vectorizes" the landmass grid by threading a mean high-water line between adjoining land and water points.[43] For regions not covered by the Landmass Blanking array, Defense Department cartographers captured

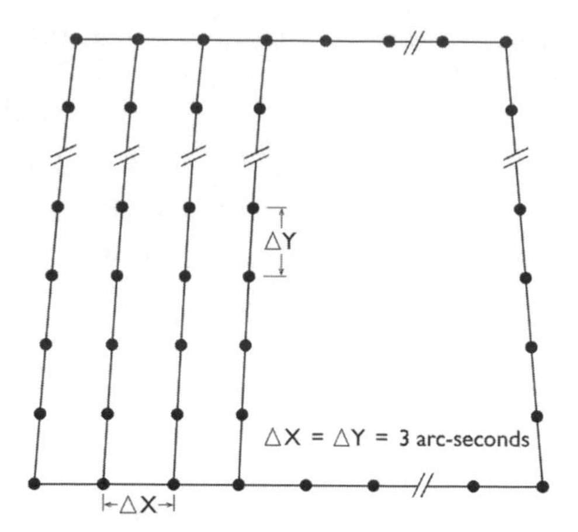

Figure 7.6. The Digital Landmass Blanking Data records the occurrence of land or water at intersection points in a grid of meridians and parallels spaced 3 arc-seconds apart.

the coastlines of existing maps no less detailed than 1:250,000, the "nominal scale" of the WVS dataset.[44] An expanded version called the World Vector Shoreline Plus includes five additional "libraries," generalized for use at 1:1,000,000, 1:3,000,000, 1:12,000,000, 1:40,000,000, and 1:120,000,000, the latter roughly similar in detail to a page-size map in a world atlas.

Are the data reliable? It depends on what you need them for. If you're writing a book on coastlines with examples focused on Five Islands, Maine, an apt answer is "not very"—especially after searching in vain for the Five Islands' landward neighbor, Georgetown Island, over eighteen square miles in area and home to 900 year-round residents. Compare figure 7.7, an extract of the 1:250,000 WVS+ library centered on Five Islands, with figure 7.8, a snapshot of the same area scanned from a U.S. Geological Survey 1:250,000-scale topographic map. Look closely at the latter, and you'll see near its center a tiny circle representing the settlement of Five Islands, appropriately labeled. Directly to its left and also labeled is the much larger Georgetown Island, mysteriously omitted from the vector shoreline plot.[45] Not surprisingly, the 1:1,000,000 WVS+ library, derived automatically from its large-scale counterpart, perpetuated an error unconscionable to the human compilers who attentively included the island on the millionth maps published in 1912 and 1957 (fig. 7.3).

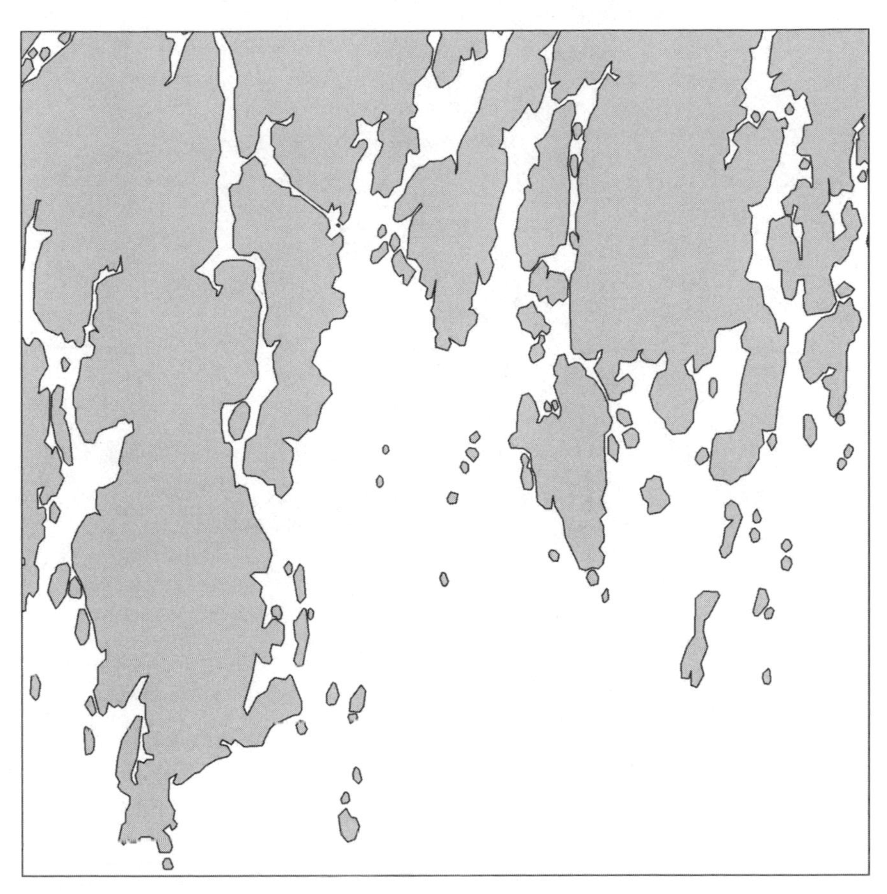

Figure 7.7. Section of the Maine coast as represented in the 1:250,000 library of the World Vector Shoreline Plus. Centered on Five Islands, this excerpt was plotted at 1:250,000. Note the absence of Georgetown Island, highlighted in figure 7.8.

The case of the missing island bolsters the argument that good cartography is seldom cheap and rarely happens overnight. The cost and complexity of compiling a worldwide cartographic database, especially for intermediate scales at which an eighteen-square-mile island might matter, encourages shortcuts such as relying on readily available data collected for an altogether different purpose. And because good cartography often depends on users who find and report errors, global shorelines in specialized databases are especially vulnerable when mariners and coastal scientists, like NOAA clients in the United States, have better data and little need for less-detailed

Figure 7.8. Excerpt from the Bath, Maine, sheet of the U.S. Geological Survey's 1:250,000-scale topographic map, published in 1963 and covering the same area as figure 7.7. The shoreline of Georgetown Island is highlighted for emphasis.

worldwide shorelines. By contrast, omission of Georgetown Island seems less likely on a general-purpose map that integrates coastlines with transport routes, boundaries, rivers, place names, and other reference features requiring greater editorial scrutiny. If the map editor doesn't catch an error, chances are good that a user will.

While a standardized, general-purpose global map seems a good idea, getting international collaborators to agree on a uniform scale is far from easy, especially now that many nations have full 1:100,000-scale coverage or better, which undermines the round-number mystique the millionth map enjoyed a century ago. There's no shortage of solutions, which range from the collaborative Global Map project proposed in the early 1990s by Japan's Geographical Survey Institute,[46] to VMAP-1, an upgraded, 1:250,000-scale version of the Defense Department's Digital Chart of the World.[47] (Developed in the 1990s with a "nominal scale" of 1:1,000,000, the Digital Chart of the World was the conceptual offspring of the Central Intelligence Agency's less detailed World Data Bank II, released in 1975 as an enhancement of the CIA's World Data Bank I, a pioneering mid-1960s effort that stimulated considerable interest in computer cartography.[48]) Other suggestions include a worldwide repository of medium-resolution orthorectified satellite imagery, referenced to a global datum and similar in concept to Google Earth, and a call for all national mapping organizations to ship a set of their latest 1:250,000-scale maps and all subsequent revisions to a private company that agrees to scan the maps, place the images on a common projection, and distribute the information on low-cost CD-ROMs or advertiser-supported Web sites.[49] (The original suggestion nominated Microsoft, but Google now seems a more likely moderator.) Whatever role the Internet assumes, users must be wary of inconsistencies that could corrupt details like small coastal islands, which can matter greatly as my next chapter illustrates.

EIGHT

BASELINES AND OFFSHORE BORDERS

Ronald Reagan's death in mid-2004 at age ninety-three unleashed a barrage of ready-to-go obituaries celebrating his heroic journey from Hollywood to the White House by way of the California governor's mansion. Distracted by the former film star's capstone role as Victorious Cold Warrior, Reagan's eulogists made no mention of Presidential Proclamation 5928, signed twenty-four days before the end of his second term—a puzzling omission insofar as the Reagan Proclamation added perhaps 100,000 square miles to the American homeland without military force or clandestine diplomacy.[1] In a single pen-stroke, our fortieth president pushed the nation's maritime boundary nine nautical miles beyond the traditional three-mile limit, anchored cartographically at "the baselines of the United States determined in accordance with international law."[2] Within its territorial sea, a country enjoys full sovereignty, including the right to enforce laws, levy taxes, and exclude foreign vessels not pursuing expeditious "innocent passage" along the coast or into a port.[3] Coastal nations can also claim legal and regulatory rights within a "contiguous zone," extending outward an additional twelve nautical miles.

The low-water shoreline, which anchors the territorial sea, is also the landward edge of a country's exclusive economic zone (EEZ), which extends outward two hundred nautical miles (370 kilometers)—unless it encounters another nation's EEZ along a line equidistant from their opposing shorelines. Within its EEZ a nation has sovereign rights to manage fisheries, mine the seabed, and extract oil and natural gas, but others can cross the area freely by sea or air as well as lay submarine cables or pipelines. Although the two-hundred-mile strip usually includes the adjacent continental shelf, a country can claim additional seabed where the shelf is broader, but the water above remains international.[4] Beyond the EEZ all nations can fish and navigate the High Seas.

Vaguely rooted in submarine topography, the exclusive economic zone is a creation of the United Nations, which became concerned about conflict over seabed exploration and fishing privileges after the United States claimed mineral rights on the adjacent continental shelf in 1945 and Chile, Ecuador, and Peru asserted exclusive fishing rights within two hundred miles of their shores in 1952.[5] Three international conferences, the first in 1958, led to the 1982 UN Convention on the Law of the Sea, which standardized the twelve-mile territorial sea and the two-hundred-mile EEZ. The United States refused to ratify the treaty because of perceived limitations on seabed mining but adopted most of its provisions, including a two-hundred-mile EEZ, proclaimed by Reagan in 1983.[6] Encompassing 3.4 million square miles, the EEZ nearly doubled the country's size and made the United States a maritime neighbor of Japan and New Zealand.[7]

In this context, Reagan's widening of the territorial sea in 1988 was little more than a postscript, proclaimed well after the rest of the world had abandoned the obsolete three-mile limit, a seventeenth-century notion based on the range of a cannon fired from the shore. A further catch-up occurred in 1999, when Bill Clinton, late in his second term, officially extended the contiguous zone outward to twenty-four nautical miles from its previous, twelve-mile position, rendered inconsequential by the Reagan Proclamation more than a decade earlier.[8]

Equally overdue was American ratification of the UN Convention on the Law of the Sea, approved by 155 other countries between 1982 and 2007, despite warnings from the Bush Administration that Russia might push its EEZ boundary northward on the Arctic shelf, into an area the United States might well claim once it signed the UN accord.[9] A key concern was the looming 2009 deadline for submitting claims to the UN Commission on the Limits of the Continental Shelf, empowered to divvy up potentially valuable mineral rights.[10] By UN rules, only signatories could file claims. Even though the United States is notorious for snubbing international accords, timely cooperation can avoid nasty confrontations later on. As of this writing, Congress had yet to endorse the agreement.

With much at stake if predictions of hidden wealth beneath the sea prove true, coastal countries cautiously delineated their new seaward boundaries and scrutinized the cartographic assertions of maritime neighbors. In theory, a maritime territory or zone is defined by a multitude of overlapping circles, twelve or two-hundred nautical miles in radius, centered at every point along the shoreline (fig. 8.1). Although an indefinitely large number of circles could be drawn, it's generally sufficient to plot a sequence of offshore arcs centered on locally prominent headlands. Delineation can be further simplified by connecting seaward points with a straight line, drawing circles at each end, and joining the circles with a straight line that just touches their circumferences (fig. 8.2). Straight baselines simplify an offshore boundary, which is most easily represented by a sequence of straight-line segments connected at so-called turning points. Especially useful for coastal Maine and other heavily indented shorelines, straight baselines can also string together offshore islands or provide a straightforward perimeter for a group of small, not-too-distant islands comprising archipelagic nations like the Republic of the Maldives, in the Indian Ocean. Archipelagic baselines (*upper left* in fig. 8.2), which can anchor vast EEZs to minute patches of dry land, epitomize the geopolitical generalization of the coastline by international law.

However convenient for marine cartographers, straight baselines that are exceptionally long can yield maritime zones outrageously broader in places than an envelope of painstakingly positioned twelve- or two-hundred-mile arcs. The problem's obvious, but there's no consensus on how long is too long. The UN Convention on the Law of the Sea, which endorses straight baselines for any coast that is "deeply indented," has "a fringe of islands . . . in its immediate vicinity," or is "highly unstable" because of "a delta or other natural conditions," cautions only that "the drawing of straight baselines must

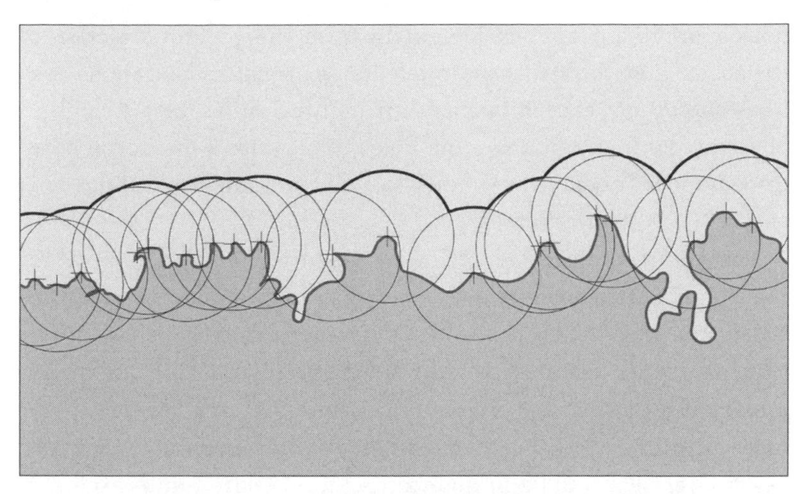

Figure 8.1. Construction of a seaward boundary as an envelope of intersecting, equal-size circles centered on forward points along a "normal" baseline.

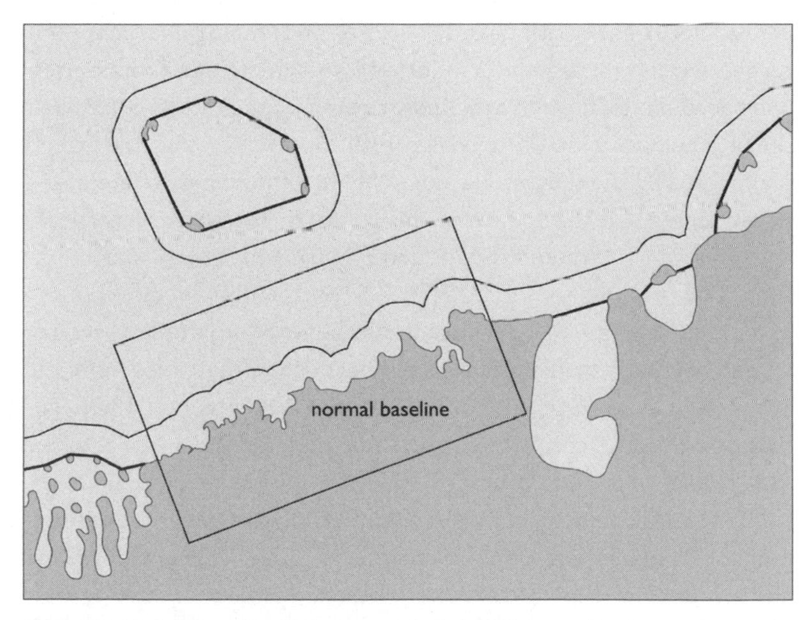

Figure 8.2. Straight baselines that span coastal indentations, enclose bays, and connect offshore islands simplify construction of seaward boundaries. The tilted rectangle highlights for comparison the "normal" baseline described in figure 8.1.

not depart to any appreciable extent from the general direction of the coast."[11] By contrast, the United States argues that straight baselines should not exceed twenty-four nautical miles, just enough to obviate potentially troublesome holes in a nation's territorial sea.[12] Unconvinced, twenty-three countries claim straight baselines longer than fifty nautical miles.[13]

Disagreements can be settled by negotiation, as in 1977, when the United States and Cuba worked out a compromise boundary between the Florida Keys and Cuba's northern coast. The two maritime neighbors had been feuding since 1959, when Fidel Castro established a Communist regime on a shoreline only seventy-seven nautical miles away. Further conflict seemed inevitable eighteen years later, when Cuba announced a two-hundred-mile exclusive economic zone and the United States claimed a similarly broad "exclusive fishing zone."[14] The State Department objected to straight baselines along Cuba's northern coast, which was "neither deeply indented nor fringed with islands."[15] After exchanging charts of their respective coasts, negotiators plotted two equidistance boundaries, one based on the normal baselines favored by the United States and the other embellishing both coasts with the straight-baseline principle favored by the Cubans. Because the two equidistance lines were never more than three nautical miles apart, the countries quickly agreed to an intermediate boundary, 313.4 miles long and represented by twenty-seven turning points, each with a precise latitude and longitude (fig. 8.3).[16] Because monuments at sea are impractical, agreed boundaries are published as lists of coordinates registered to a mutually acceptable horizontal datum. Fortunately for the negotiators, both the United States and Cuba had tied their charts to the 1927 North American Datum.

As the Florida Keys illustrate, a small island well offshore can significantly enlarge a nation's maritime territory. Without Key West and an often inconspicuous cluster of cartographic dots directly west named Dry Tortugas for their abundant sea turtles and lack of fresh water, the western portion of the U.S.–Cuban boundary would swing noticeably to the northwest, to widen Cuba's EEZ at the expense of the United States. Because international law treats islands like any other part of a country's coastline, they can dramatically influence

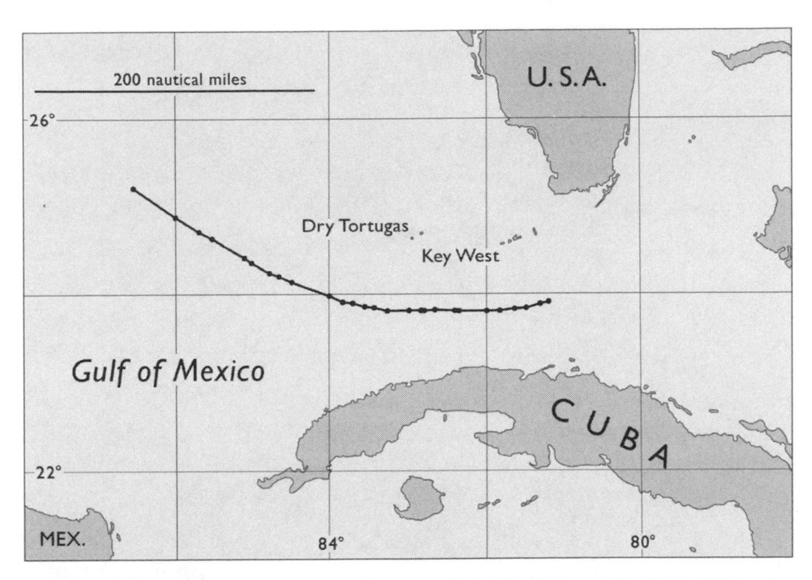

Figure 8.3. Compromise maritime boundary negotiated by Cuba and the United States in 1977. Redrawn from map in U.S. Department of State, *Limits in the Seas* no. 110, 1990.

the "median line" drawn between coastal neighbors by connecting points equidistant from their respective shorelines.

In addition to pushing a maritime boundary well beyond the mainland, an offshore island can significantly divert the seaward extension of a well-established international boundary. To viewers unfamiliar with rules, an apparent inconsistency between a landward boundary and its maritime counterpart can appear to suggest devious diplomacy or armed aggression. That's an impulsive reading of the Pacific boundary between the United States and Mexico, which turns abruptly southward from a hypothetical seaward projection of Baja California's northern boundary (fig. 8.4).[17] Although Yankee imperialism no doubt influenced the existing landward boundary, well-established American ownership of San Clemente and San Nicolas islands largely accounts for the maritime boundary fixed in 1978, when the two countries agreed to four turning points that approximate the equidistance line. Without Guadalupe, an offshore island that redirected the boundary westward, Mexico's EEZ would be even smaller.

"Provisional" maritime boundaries are not official until the "draft

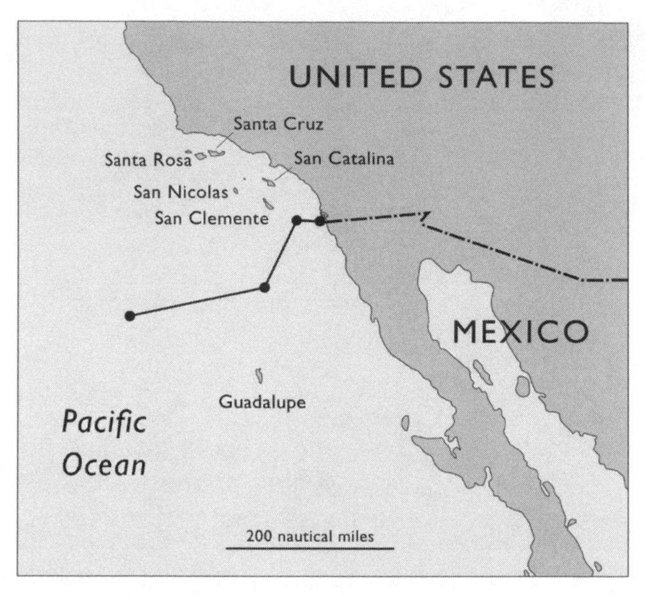

Figure 8.4. Maritime boundary in the Pacific Ocean negotiated by Mexico and the United States in 1978. Compiled from maps in Prescott and Schofield, *The Maritime Political Boundaries of the World*, 579; and U.S. Congress, House Committee on Foreign Relations, *Three Treaties Establishing Maritime Boundaries between the United States and Mexico, Venezuela, and Cuba*, 15.

treaty" is endorsed by whatever bodies have the final say. Approval can be delayed indefinitely, as when the United States Senate, which has the constitutional responsibility to approve all treaties, postponed debate on the U.S.–Mexican agreement after Hollis Hedberg, an emeritus professor of geology at Princeton University, objected to the boundary delineated in the Gulf of Mexico.[18] Hedberg, who considered submarine topography more important than offshore islands, testified that the treaty "would lose to the United States without any just or valid reason an area of deep water [with] prospective energy resources of potentially great value"—roughly 25,000 square miles, he figured.[19] Not brought up for a vote, the treaty was put on the State Department's "pending" list, which did not deter the Interior Department from auctioning oil leases in parts of the Gulf where U.S. drilling rights were uncontestable.

The unsigned 1978 treaty resurfaced in 1997, when industry representatives, eager for additional leaseholds, approached the Senate

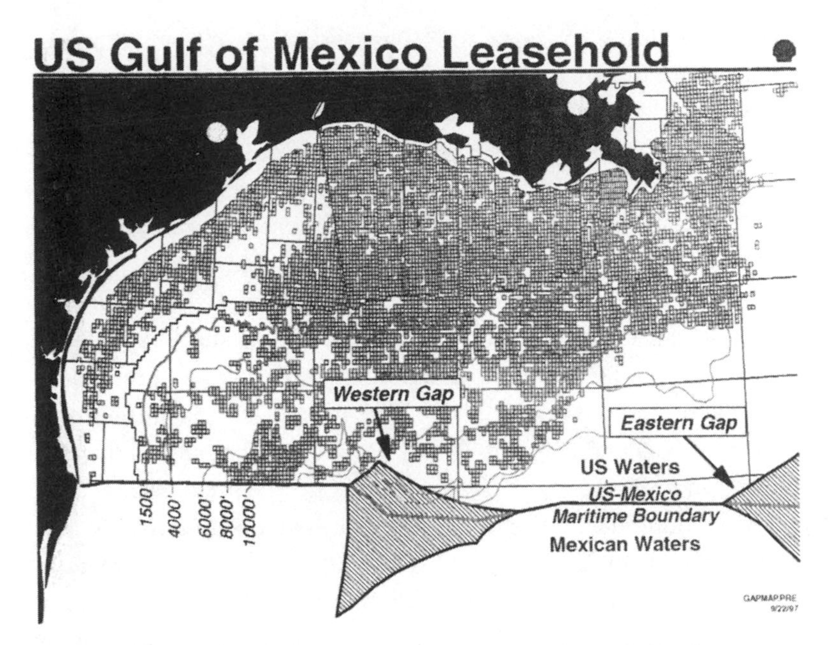

Figure 8.5. Gaps between the Mexican and American EEZs in the Gulf of Mexico were deemed ripe for petroleum extraction if the United States and Mexico were to divvy them up with an equidistance line. From U.S. Congress, Senate Committee on Foreign Relations, *U.S.–Mexico Treaty on Maritime Boundaries*, 11.

Committee on Foreign Relations about two "donut holes" in the Gulf between the American and Mexican EEZs (fig. 8.5).[20] Recent exploration suggested the Western Gap might be especially productive, and possibly rival oilfields in the Persian Gulf. The two nations could legally divvy up the two gaps, but Mexico refused to negotiate until the Senate ratified the 1978 treaty. The State Department's list of treaties in force reports an outcome that strongly suggests "it's all about oil":

> Treaty on maritime boundaries. Signed at Mexico City May 4, 1978; entered into force November 13, 1997.
> Treaty on the delimitation of the continental shelf in the western Gulf of Mexico beyond 200 nautical miles, with annexes. Signed at Washington June 9, 2000; entered into force January 17, 2001.[21]

While coastal charts play an important role in fixing baselines and drawing maritime boundaries, the Law of the Sea is far more complex than the concepts and rules I've cited.[22] The equidistance principle, for instance, is not the only basis for splitting up intervening waters or extending landward boundaries seaward. Neighboring states can extend a seaward boundary perpendicular to the coast or anchor it to a meridian or parallel. International tribunals often consider historic claims, such as traditional fishing grounds, more important than baselines and geometry. Submarine topography, especially shallow waters that suggest continuity with mainland territory, can claim portions of the continental shelf beyond the two-hundred-mile boundary as well as enlarge a nation's territorial sea. Islands, characterized by dry land at high tide, have their own territorial seas and EEZs, while rocks "which cannot sustain human habitation or economic life of their own" do not.[23] Even so, a "low-tide elevation," defined as "a naturally formed area of land which is surrounded by and above water at low-tide but submerged at high-tide" can be used to extend the baseline for a country's territorial sea as long as it lies within the territorial sea otherwise defined by the mainland or a legitimate island.[24]

Closing lines that rope off huge portions of the ocean are inherently contentious, particularly where a questionably expansive territorial sea would exclude a major power's merchant vessels, warships, and aircraft. International law allows these claims if a country can demonstrate a well-publicized, effectively enforced, long-standing, and continuous assertion of sovereignty generally respected by other countries.[25] According to the U.S. Supreme Court, the waters of Long Island Sound and Mississippi Sound satisfy the criteria but Nantucket Sound, Block Island Sound, Florida Bay, Santa Monica Bay, San Pedro Bay, and various indentations along the Louisiana coast did not.[26] Not all nations are similarly circumspect. A 1992 State Department study lists seventeen "excessive" claims to historic bays that the United States has either protested or defied openly by entering the area with warships or military aircraft to assert its right to unrestricted travel.[27] Understandably, the "excessive" claims of traditional allies like Australia, Canada, Italy, and Portugal were not challenged with a show of force.

To bolster its argument, the State Department described some of the more egregious "historic" claims with maps. "Libya's Claim to the

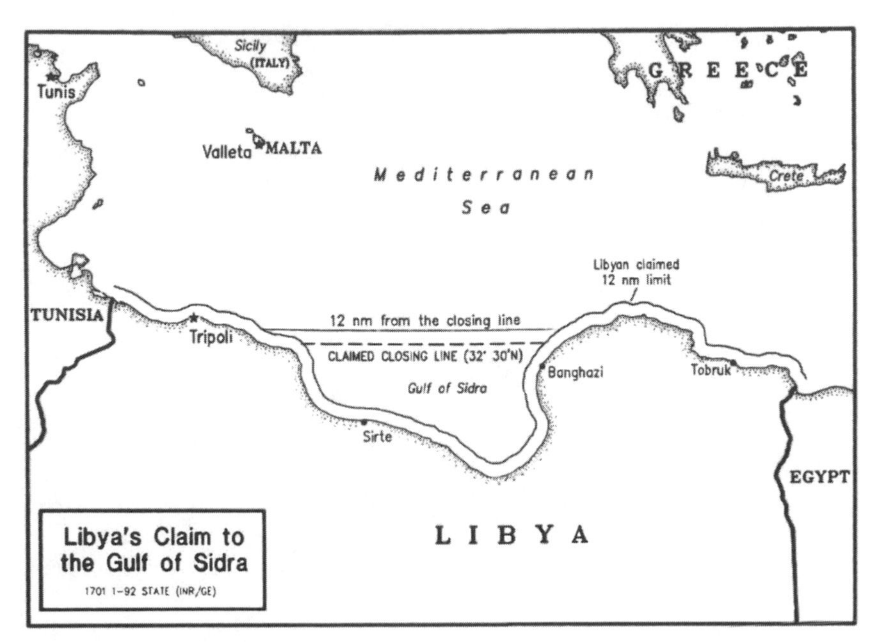

Figure 8.6. Libya's Claim to the Gulf of Sidra. From U.S. Department of State, *Limits in the Seas* no. 112, 1992, map 4.

Gulf of Sidra," reproduced here as figure 8.6, shows a closing line anchored by the parallel at 32° 30′ N and "approximately 300 miles long"—far longer than the 24 nautical miles allowed by the UN Convention on the Law of the Sea.[28] The United States protested in 1974, a year after the Libyan Foreign Ministry announced the claim, and again in 1979 and 1985. Underscoring American objections to "unlawful interference with the freedoms of navigation and overflight,"[29] the Defense Department has "conducted operational assertions" on numerous occasions since 1981.[30] France and Spain also filed protests, as did three non-Mediterranean nations, Australia, Norway, and West Germany. Although complaints and armed assertions have not persuaded Libya to back down, its enlarged territorial sea seems little more than a line on a map.

While questionable restrictions on international waters provoke diplomatic protests and military challenges, overlapping seabed and fishing claims are fodder for the International Court of Justice, located in The Hague, where Canada and the United States argued

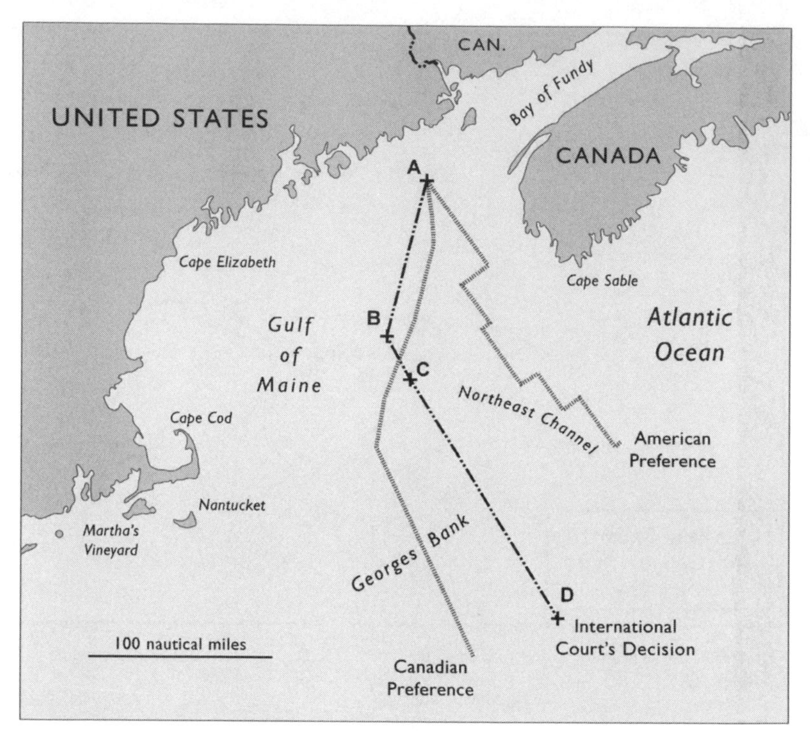

Figure 8.7. Maritime boundary between Canada and the United States in the Gulf of Maine is closer to the position favored by Canada. Compiled from maps in International Court of Justice. "Case Concerning Delimitation of the Maritime Boundary in the Gulf of Maine Area," 289, 346.

over their common boundary across the Gulf of Maine.[31] To simplify debate, the two countries agreed in 1981 that the boundary should begin at a point near the mouth of the Bay of Fundy (point A in fig. 8.7). Canada favored an equidistance line based on a generalized coastline that ignored Cape Cod, Nantucket, and Martha's Vineyard, while the United States contended that its EEZ should reflect the configuration of the continental shelf and include the Northeast Channel as well as Georges Banks, a fishing ground "as American as apple pie."[32] After reviewing "some 7,600 pages of pleadings and 2,000 pages of oral arguments together with 300 supporting maps, sketches or diagrams," the court announced its decision in October 1984.[33] A compromise boundary described by only four turning points rejected Canada's equidistance argument and the United

States' assertion that Georges Bank was an indivisible ecological unit. The first segment (A–B) bisects the angle between straight lines converging on the starting point from Cape Sable and Cape Elizabeth, and its other sections are median lines between Nova Scotia and Massachusetts. Although it's tempting to suggest the judges split the difference between the two positions, the decision line more closely matches the Canadian delineation.[34]

In addition to defending its claims before international tribunals, the federal government also fights off coastal states eager to extend their individual maritime boundaries. Although the Reagan Proclamation widened the territorial sea from three to twelve nautical miles, state lands still stopped at the old three-mile limit, except for Texas and Florida, which entered the Union with a three-league (nine-nautical-mile) maritime belt in the Gulf of Mexico.[35] Considerable litigation followed the enactment of the Submerged Lands Act in 1953, when Congress gave coastal states full ownership of lands beneath navigable waters within their boundaries. Among the earliest cases was Louisiana's assertion that it too was entitled to a three-league sea. Because the Supreme Court found the claim unconvincing, the map of federal petroleum leases (fig. 8.5) shows a noticeably wider buffer along the Texas shoreline.

More recently Alaska petitioned the high court for a wider band of seabed along its oil-rich Arctic shoreline. The state's argument focused partly on Dinkum Sands, a would-be island northeast of Prudhoe Bay more than three miles from the shore. Named after the *Fair Dinkum*, the boat of the Coast and Geodetic Survey party that discovered it in 1949, Dinkum Sands is a formation of gravel and ice less than a quarter-mile long.[36] Covered by pack ice much of the year, it slumps below the mean high-water level in late summer, when the ice melts. The only elevation measurement made during winter found Dinkum Sands below mean high water, which suggests it's merely a low-tide elevation too far from the shore to enhance the state's territorial sea.[37] Even so, Alaska insisted the feature is indeed an island, and thus entitled to its own three-mile sea, which would yield to the state additional oil revenue. In a 1996 decision, the U.S. Supreme Court rejected the claim as well as a back-up argument that Dinkum Sands was an island at least part of the year. "Quite

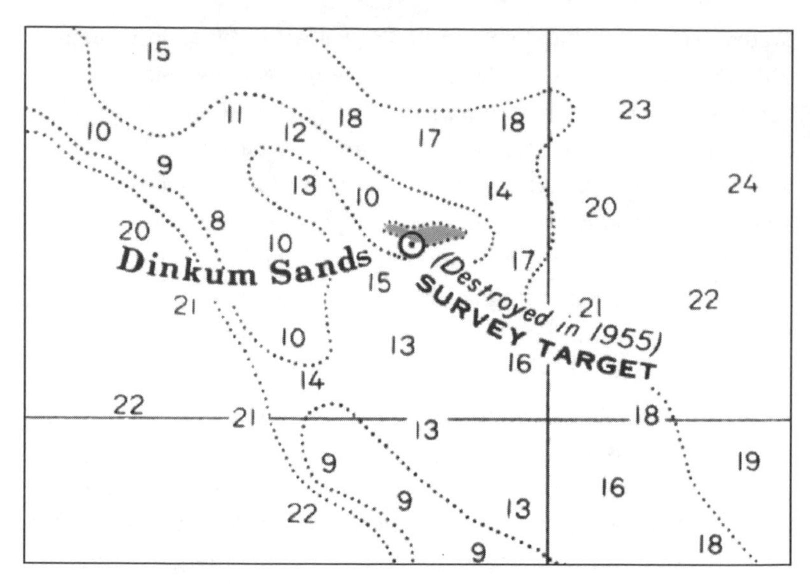

Figure 8.8. Dinkum Sands, Alaska, as shown on the May 1974 edition of U.S. Coast and Geodetic Survey chart 16046, McClure and Stockton Islands and Vicinity, published at 1:50,204. According to the chart's scale bar, the "island" shown here is slightly shorter than 0.2 nautical miles.

apart from the fact that Alaska's proposal would lead to costly and time-consuming monitoring efforts," Justice Sandra Day O'Connor wrote for the court, "no precedent [exists] for treating as an island a feature that oscillates above and below mean high water."[38]

Dinkum Sands owes its notoriety to Harley Nygren, a member of the 1949 survey team and a uniformed officer in the Coast and Geodetic Survey.[39] (Currently called the NOAA Corps, this cadre of commissioned officers traces its origin to 1807, when Congress founded the Survey of the Coast.) Nygren's survey put Dinkum Sands on the map and kept it there, even after a U.S. Navy ship that visited the area in 1955 could find neither the island nor the vertical marker ("survey target" in fig. 8.8) that Ensign Nygren had erected six years earlier.[40] In the 1970s, when Rear Admiral Nygren was a member of the interagency U.S. Coastline Committee, he convinced other members that Dinkum Sands was more than the low-tide elevation depicted on the nation's nautical charts after 1955.[41] The committee gave the feature a three-mile sea, which subsequently appeared on

1979 mineral leasing maps as state land. The case first went to court in 1979, after an Arctic expert in the Interior Department questioned the Coastline Committee's interpretation and the federal government moved to erase the boundary.

Dinkum Sands and the other boundary disputes examined here illustrate a legalistic, rule-based form of coastline generalization with outcomes not radically different from the mapmaker's practice of smoothing out irregular shorelines on small-scale maps and amalgamating small islands with the mainland. Although cartography and the Law of the Sea are different in their techniques and impact, both approaches reflect the geometric complexity of a natural feature more easily appreciated in a simplified form.

NINE

CALIBRATING CATASTROPHE

High- and low-water shorelines on nautical charts are not the only significant land-water boundaries. A third shoreline shown only on specialized maps affects the life and livelihood of most coastal residents. These maps include Flood Insurance Rate Maps, FIRMs for short, which portray the likely impact of rare yet inevitable storms that carry surf and seawater well inland, destroying homes and blocking evacuation routes. (For an example, see figure 2.6, from the FIRM for Five Islands, Maine.) This third shoreline is often called the 100-year flood line because it marks the extent of flooding caused by a storm with a "recurrence interval" of a hundred years—a potentially misleading expression suggesting that floods this extensive occur a hundred years apart. What's really shown is an inundation with a one-percent chance of happening in any given year. If you suffered a 100-year flood last year, there's still a one-in-a-hundred chance of a similar disaster this year. And if you live within, or even near, the flood zone, you'd better have flood insurance as well as an evacuation plan that includes heirlooms and family pets.

The 100-year flood line is the focus of two stories: one about how mapmakers plot its position and the other describing its use in cutting losses and planning evacuations. The first story begins by conceding that the 100-year flood line is arbitrary. Although severe coastal storms are irrefutably real—ask New Orleans residents displaced by Hurricane Katrina in 2005—the 100-year flood zone defies exact delineation because of unavoidable uncertainty in climatological prediction and numerical modeling. Living on the dry side of its inherently vague boundary is no excuse for complacency. Even if the 100-year flood line could be plotted precisely, a less frequent 500-year flood could devastate a much wider area. A hundred years is merely a convenient, credibly long timeframe for humans who seldom live past their nineties and prefer round numbers. If each hand had an extra digit, we'd probably be planning for a 144-year flood.[1]

Federal officials and local communities tolerate this arbitrariness because flood insurance is unworkable without a rational, systematic basis for charging higher premiums in riskier places. With its scientific shortcomings understood but overlooked, the FIRM is an excellent example of what theorists in cultural studies call a social construction.[2] Flood mapping is similar to economic theory and organized religion in that flood lines become real because they are useful. Coastal-zone managers, who have little use for social theory and postmodern jargon, think of the FIRM as a flawed planning tool that is nonetheless valuable in protecting homeowners from devastating losses.[3] The maps aren't perfect, but to qualify for subsidized flood insurance, communities must require contractors to raise the lowest floor of new buildings to the 100-year flood elevation, or higher.[4]

Coastal flooding is not the only concern of mapmakers at the Federal Emergency Management Agency (FEMA), which runs the National Flood Insurance Program. Riverine flooding along sluggish streams as well as the mighty Mississippi also requires flood maps. Because coastal and riverine flooding reflect diverse physical processes and yield dissimilar data, the hydrologists and civil engineers who delineate flood zones must treat these situations differently even though their flood maps are similar in appearance and objectives.

Risk maps for riverine flooding rely on stream gauges and topographic maps, both of which were available when the federal government began mapping flood zones in the 1970s.[5] A continuous, day-by-day record of the peak per-second volume of water flowing past a particular point during the last thirty or more years provides a statistical basis for predicting extreme flooding at the gauging station itself as well as for areas nearby. Because extrapolating from 30 years of data to a 100-year recurrence interval seems within the limits of mathematical conjecture, hydrologists had few qualms about estimating the additional height of water in the channel during a 100-year flood and using contour maps to determine the area underwater. For example, if the normal elevation of water in a river is 420 feet and the 100-year flood would raise the level 10 feet, the flood boundary would coincide with the 430-foot contour. Additional uncertainty arises when the mapmaker must interpolate between gauging stations or guess the configuration of intermediate contours not shown on the map. Overgeneralized or out-of-date elevation maps further undermine reliability, as do climate change, destruction of wetlands, and metropolitan development that accelerates runoff by covering the land with rooftops, roads, and parking lots. As one hydrologist described the process, mapping flood zones is like to trying to hit "a moving target."[6]

Flood mapping in coastal environments is more involved than its riverine counterpart because water-level data collected by tide gauges don't reflect the full complexity of catastrophic storms, submarine topography, irregular shorelines, and coastal erosion. Hydrologists can interpolate reliably between gauging stations along a river if the floodplain is relatively homogeneous and intervening tributaries are evenly spaced.[7] Straight stretches of coastline with a uniformly sloping seafloor might allow simple interpolation, but irregularities in either the shoreline or underwater terrain can yield marked differences in storm surge, the name for the increased water level built up by the intense surface winds of a tropical storm or nor'easter. Although extremely low barometric pressure near the storm's center adds marginally to the water level, most of the rise—on top of whatever a high tide might create—results when strong winds push seawater onto the land. Storm surge is generally greater where the seafloor slopes more gently away from the shoreline, but an uneven seafloor and irregular coastline can aggravate the pile-up in some places and dampen it in others. Surge heights of twelve feet are not uncommon during a category 3 hurricane (winds 111 to 130 mph), and a category 5 storm (winds greater than 155 mph) can add a surge of eighteen feet or more—especially

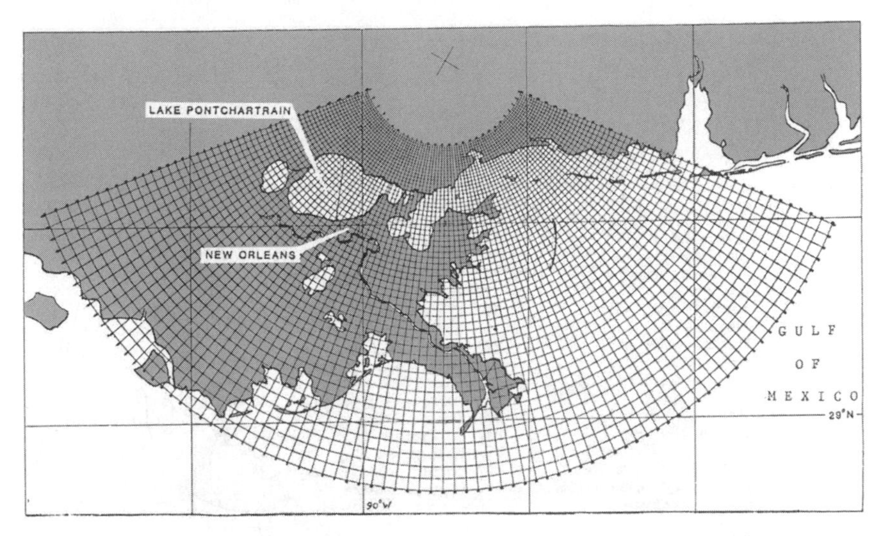

Figure 9.1. SLOSH basin for New Orleans, Lake Pontchartrain, and the Mississippi delta. From Jelesnianski, Chen, and Shaffer, *SLOSH: Sea, Lake, and Overland Surges from Hurricanes*, 30. Gray shading was added to indicate land areas.

devastating if it occurs at high tide when the sun and moon are aligned.[8] Wave action adds to the increase in "stillwater level," and because winds in the northern hemisphere move counterclockwise around the storm's center, places to the right of the storm path generally experience the greatest surge shortly before the storm center moves inland.

Numerical representations of the coastline and seafloor help coastal scientists forecast the local impact of surge from a hypothetical storm with a specific path, speed, and pattern of barometric pressure around its center. Because the modeling technique is called SLOSH, an acronym for Sea, Lake, and Overland Surges from Hurricanes, these local databases are known as SLOSH basins.[9] As figure 9.1 illustrates for New Orleans and the Mississippi delta, a SLOSH basin is a circular mesh anchored to an interior point chosen to provide balanced coverage around a specific harbor or city. The polar grid yields appropriately smaller cells toward the center of the basin, where finer resolution is needed, and also lets the computer calculate each cell's area as well as its distance from the storm center. Offshore cells have negative elevations, registered seamlessly to

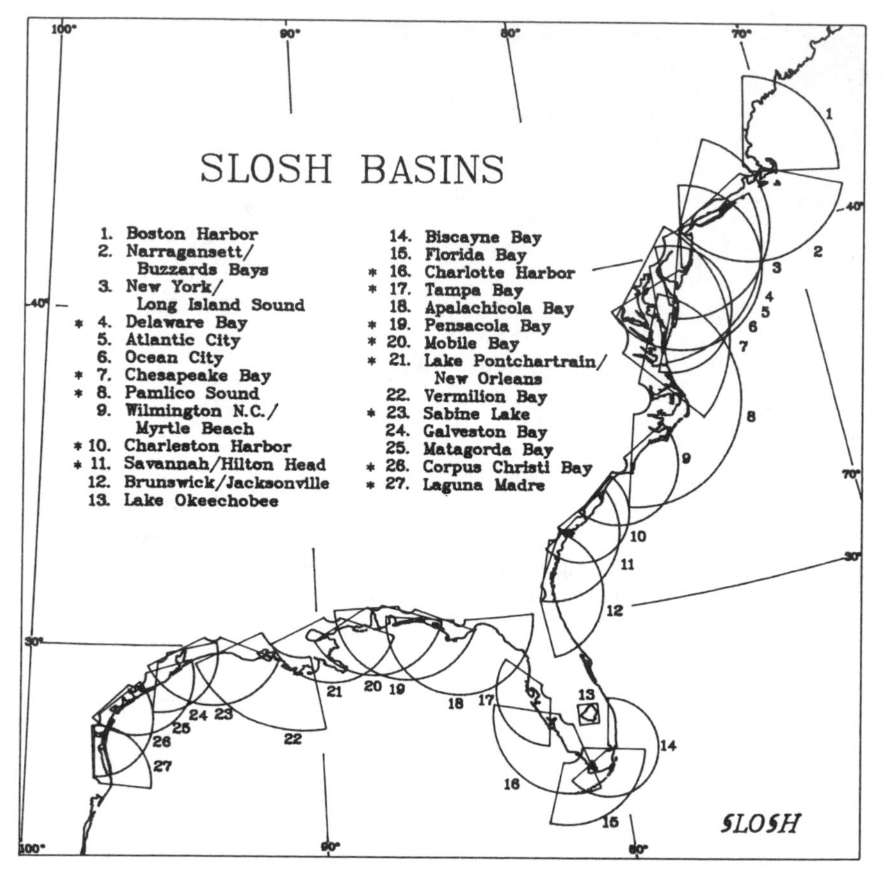

Figure 9.2. SLOSH basins developed in the 1970s and 1980s for the Atlantic and Gulf coasts. From Jelesnianski, Chen, and Shaffer, *SLOSH: Sea, Lake, and Overland Surges from Hurricanes*, 64.

the same vertical datum (see pp. 17–18) as their landward counterparts. The computer can calculate the volume of water in a cell by multiplying its area by the elevation difference between the water surface and the seafloor or land surface below.

To encourage surge modeling, the National Weather Service originally developed twenty-seven SLOSH basins centered on Atlantic and Gulf coast locations between Boston and Laguna Madre, near the Mexican border. As figure 9.2 shows, overlap is needed to provide accurate assessments for comparatively close coastal cities like Mobile and Pensacola. Note the SLOSH basin for Lake Okeechobee (no. 13), which is large enough for hurricane winds to whip up a destructive surge along its shoreline. In 1928, a hurricane that pushed out of the lake killed over 2,500 people, and recent studies indicate an even greater catastrophe is possible today.[10] There are now forty SLOSH basins, which afford denser coverage of the Gulf Coast as well as new areas such as Oahu, Hawaii; Puerto Rico; and Penobscot Bay, Maine.[11]

SLOSH models operate like a system of bank accounts set up to allow transfers of water across shared boundaries between neighboring cells. Inland cells begin with a zero balance, offshore cells start off with a positive balance, and new water enters the system from the sea beyond the outer band of cells. Although the simulation involves a lot of number crunching, the process is straightforward. If more waters flows into a cell than flows out, its water level rises. Surge occurs when an inland cell, which started with nothing, gains a positive balance. The rules are simple: no cell can overdraw its account, and for every transaction across a cell boundary, the debit on one side must match the credit on the other.

Transfers from a cell to its neighbors are controlled by wind friction, which the model calculates from the barometric pressure at the storm's center, the change in pressure outward from the storm center, the cell's distance and direction from that center, and in more advanced models, the aerodynamic drag of marsh grass and buildings. Because circulation is counterclockwise around the storm's center, surge-producing winds are typically greater (as noted earlier) to the right of the storm track just before the center hits land.[12] Opposing forces include inland elevations too high for easy inundation, and gravity, which resists a water elevation much greater than in neighboring cells. Flooding occurs when a cell above the shoreline accumulates surplus water, and if the surge is sufficiently strong, some of this surplus will spill into neighboring cells. The model moves forward in time in short steps—every 100 seconds, say[13]—and each step involves several rounds of adjustment to account for change in the wind field as the storm advances, as well as a succession of local adjustments needed when a cell must pass some of its new water along to neighbors immediately inland, which in turn surrender some of their gain to adjacent cells still farther

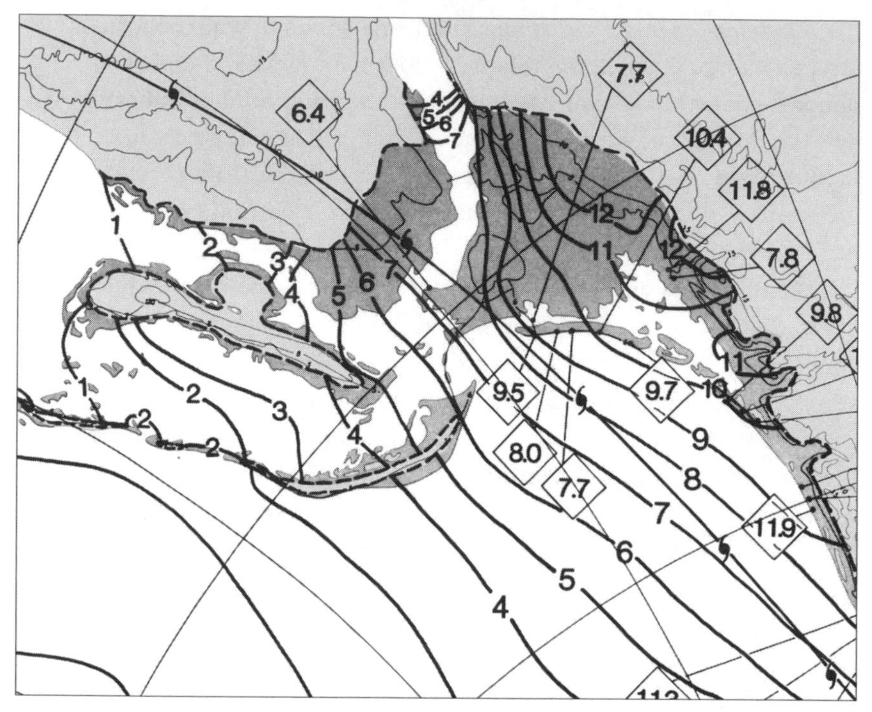

Figure 9.3. Excerpt from a simulated storm-surge map for the Charlotte Harbor, Florida SLOSH basin based on Hurricane Donna, 1960. From Jelesnianski, Chen, and Shaffer, *SLOSH: Sea, Lake, and Overland Surges from Hurricanes,* 51. Bold dashed lines show the predicted reach of storm surge. Light gray shading was added to clarify other land areas.

inland. More sophisticated approaches combine surge simulations with models of wave energy.[14] Further improvements recognize additional contributions from inland rain, especially troublesome with slow-moving storms.

By tracking the highest water level in each cell, the surge model yields a map like the excerpt in figure 9.3, a simulation of Hurricane Donna, a category 4 storm that hit Charlotte Harbor, Florida, from the southeast in 1960.[15] Symbols resembling an overlapping 6 and 9 identify the hurricane track, and the numbers on the thick black contour lines depict the water's maximum rise, in feet. Note that this "surge envelope" is thicker to the right of the storm track. Dark gray shading indicates flooding, which occurs wherever the water is higher than the land. This simulation was a "hindcast" run to test the model using estimates of Donna's intensity and trajec-

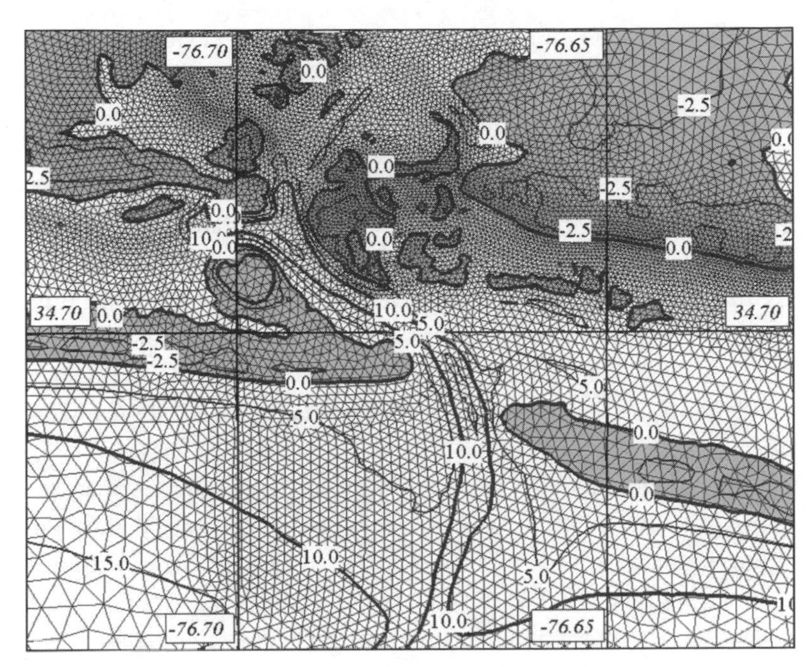

Figure 9.4. Portion of an unstructured flood modeling grid for Beaufort Inlet, North Carolina. Vertical and horizontal lines are meridians and parallels, with longitude or latitude attached. Because labels astride contour lines indicate depth below the vertical datum, negative numbers represent elevations above the water line. From Feyen and others, "Development of a Continuous Bathymetric/Topographic Unstructured Coastal Flooding Model to Study Sea Level Rise in North Carolina," 351.

tory. The diamond shaped symbols, which show actual high-water marks measured after the storm, indicate that the computer model provided a useful but not exact prediction of local flooding. Close enough, the National Weather Service believes, for identifying evacuation zones.

Newer models use unstructured grids like that in figure 9.4 to account for irregular shorelines and small features on- and offshore, which can accelerate or retard storm surge.[16] While the coarse polar grid of the SLOSH basin in figure 9.1 cannot adequately reflect conditions along its flanks, where cells as wide as 4 miles (7 kilometers) undermine accuracy, a fine-grained unstructured grid can cover complex low-relief areas with cells as narrow as 160 feet (50 meters).[17] By taking advantage of bigger, faster computers and more refined elevation data from lidar, and by incorporating the effects of wave action and tidal dynamics, these newer models let coastal scientists and emergency management officials experiment with a more diverse range of hypothetical storms and plausible landfall locations, and make evacuation forecasts more reliable.[18] Reliable models are important not only because people can drown but also because an unnecessary exodus from comparatively safe areas will delay evacuation from areas likely to be inundated.[19] According to Stephen Leatherman, director of the International Hurricane Research Center at Florida International University, SLOSH "has served the nation well for three decades, but it's old technology now."[20] A polite way, perhaps, of calling the Weather Service warhorse a dinosaur ready for retirement.

Getting from a single simulation to a flood-insurance map requires historical data on storm frequency and intensity, multiple surge simulations, and more subjectivity than the FIRM's crisp lines seem to suggest. I looked for a clear-cut description of the process but found none.[21] Even so, an intensive search of bibliographic databases and government Web sites turned up two revealing documents. The first is FEMA's *Guidelines and Specifications for Flood Map Production Coordination Contractors,* a handbook for the outside contractors who actually make the maps. Published in 1999 and still in use as of this writing, the handbook contains a prototype Flood Insurance Study report for the hypothetical town of Floodport, Massachusetts, in the hypothetical Flood County, north of Boston and subject to riverine as well as coastal flooding.[22] An exemplar of numerous studies prepared for municipalities participating in the National Flood Insurance Program, the twenty-four-page model report contains a concise description of Floodport, the hydrological and hydraulic methods used to delineate its 100-year and 500-year flood boundaries, and the proper use of the flood map by citizens and local officials. Although flood boundaries are based on a "wave runup computer program" more relevant to the area than SLOSH modeling, most of the technical studies listed were published in the late 1970s.[23] As the FEMA handbook notes, "A variety of analytical methodologies may be used to establish BFEs [base flood elevations] and floodplains throughout coastal areas of the United States. These methodologies are too voluminous to be included in these Guidelines."[24] In other words, flood mapping is not a straightforward, standardized, wholly objective process. As with sausage making, you wouldn't want to watch.

If you do watch, you might reach conclusions similar to those of the National Research Council (NRC) committee that investigated FEMA's flood-zone maps for Lee County, Florida, which includes Fort Myers and Sanibel Island. County officials had requested the study because they thought FEMA's flood insurance maps overstated the risk, making insurance more costly and coastal property more difficult to sell. The committee looked carefully at FEMA's maps and the science behind them. Its report, published in 1983, is technically detailed but incisive. FEMA's hydrodynamic model, developed by an outside contractor, suffered from "an incorrect computational procedure for calculating bottom friction in flooded areas, an inappropriate treatment of the wind stress on the water in flooded areas where vegetation and other obstructions to the flow protrude through the water surface, and an inappropriate treatment of the flow of water over submerged barrier islands and through inlets."[25] What's more, the model's inputs were suspect because "it is difficult to estimate the frequency and probable characteristics of hurricanes likely to strike each portion of the U.S. coast."[26] Because the length of coastline and period of time used to compute storm probabilities were questionable, the flood maps were "biased toward intense storms."[27] The panel calculated its own probabilities, ran its own simulations, and found that FEMA's flood modeling "produced a 100-year surge at the coast approximately 15 percent higher" than a more scientifically defensible simulation.[28] Despite these flaws, the committee concluded that "the basic approach . . . is sound [and] appropriate for estimating 100-year flood elevations in communities where severe flooding is caused by hurricane storm surges."[29] See now why I call FEMA's FIRMs a social construction? Even though contemporary models are more sophisticated and reliable than those in use a quarter century ago, flood-zone mapping will never be an exact science.

Environmentalists, coastal scientists, and fiscal conservatives are leery of FIRMs because federally subsidized flood insurance takes much of the risk out of owning homes in coastal flood zones better left undeveloped.[30] Orrin Pilkey, a geologist famous for his rants against shoreline development, is particularly concerned about fragile barrier islands, where accelerated residential development is "utterly irresponsible."[31] Pilkey blames government programs like bridge construction, reconstruction assistance, and subsidized flood insurance, which gives middle-income buyers a sense of financial security, helps them get bank loans, and shifts much of the burden to federal taxpayers.

How much are flood maps to blame? In a 1982 study that recognized the maps' contribution to more rigorous building codes, the U.S. General Accounting Office (renamed the Government Accountability Office in 2004) concluded that subsidized flood insurance was only "a marginal added incentive for development."[32] Even so, a recent report suggested that more accurate, more accessible FIRMs can help reduce repeated claims as part of a revamped strategy that includes buyouts of repeatedly rebuilt buildings and "actuarially based rates," which would make owners more responsible for the cost of insuring property in high-risk locations.[33] Another concern is coastal erosion, which is not covered by federal flood insurance but difficult to disentangle from flood damage.[34] Congress is rightly concerned that RLPs, the bureaucrat's acronym for "repetitive loss properties," account for 30 percent of the pay-out but only 1 percent of the policies.[35] What's more, 63 percent of the RLP payout goes to only five states (Louisiana, Texas, Florida, North Carolina, and New Jersey), all coastal. According to a recent report by the Congressional Research Service, the questionable accuracy of FEMA flood maps, which sometimes place RLPs outside the flood zone, and thus beyond the reach of more rigorous building codes, is very much "a contributing factor for repetitive losses."[36]

Local emergency managers, who appreciate flood insurance and stringent building codes, are more concerned with evacuating people than rebuilding homes. Maps with arbitrary 100-year flood lines are far less important than worst-case scenarios and catastrophic storms, competently limned with improved high-resolution storm-surge models. Equally important are hurricane evacuation models, which integrate surge simulations with depictions of populations at risk and the road network. Planners seeking reliable evacuation

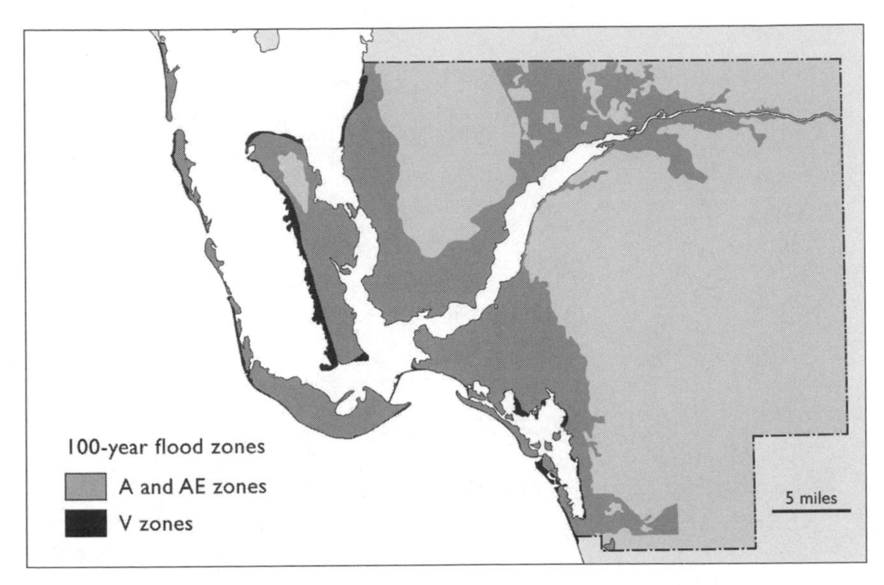

Figure 9.5. Generalized 100-year flood area in Lee County, Florida. Adapted from Lee County, Florida, GIS Department, Lee County Flood Zones.

routes that minimize congestion and avoid choke points—low-lying areas that flood early—must consider a range of plausible storms, varying in intensity, direction, and rate of advance. Their maps need to show well-known roads and other landmarks useful in helping local residents understand quickly who's endangered and how to get out.

I found a revealing example in Lee County, Florida, which is more wary of coastal flooding today than in the early 1980s. Its Web site serves up several storm-hazard maps, three of which illustrate dramatically that the third coastline is difficult to define, much less delineate. Figure 9.5, is a greatly reduced, black-and-white rendering of the county's generalized flood map, which shows six different zones in contrasting colors.[37] Except for scattered areas in the north, most of Lee County lies in a 100- or 500-year flood zone. My condensed map highlights the A, AE, and V zones, which collectively define the 100-year flood line with comparative certainty.[38] Base flood elevations, used for insurance and regulatory purposes, are defined for the AE zone (the E stands for elevation), and more restrictive land-use controls apply in the V (velocity) zone, singled out as particularly

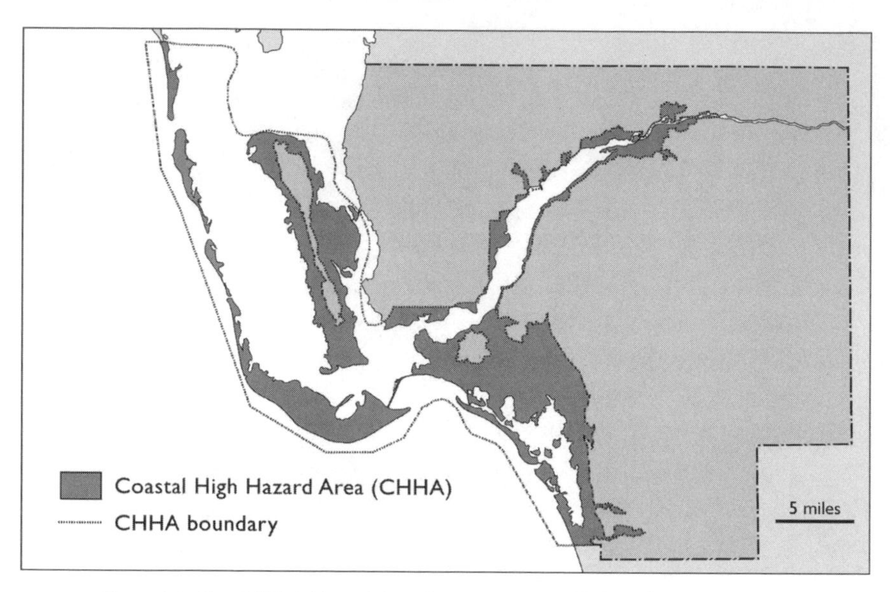

Figure 9.6. Coastal High Hazard Area for Lee County, Florida. Adapted from Lee County, Florida, Department of Community Development, Lee County Coastal High Hazard Area.

vulnerable to wave action and erosion. Note that my composite 100-year flood zone covers low-lying islands, extends more than a mile inland in most places, and includes low-lying areas farther inland near the river, which readily overflows during intense, prolonged rainstorms.

My second map (fig. 9.6) depicts Lee County's Coastal High Hazard Area (CHHA), a model-based evacuation zone for category 1 hurricanes (winds 74 to 95 mph). Delineated by the Southwest Florida Regional Planning Council for Lee and neighboring counties, the CHHA is a regulatory ploy to keep housing density in check, limit public funding of roads and sewer lines in high-risk areas, and minimize the number of people who might need to evacuate.[39] Much wider than the V zone in figure 9.5 but largely inside the A zones, the CHHA defines a less alarmist third coastline. Like other hypothetical inundation areas, the CHHA is subject to periodic revision as new data and techniques emerge. In 2006 a study commission appointed by Florida's governor to review the state's high-hazard zones found surge modeling an appropriate and effective planning technique but called for more refined elevation data, based on lidar.[40]

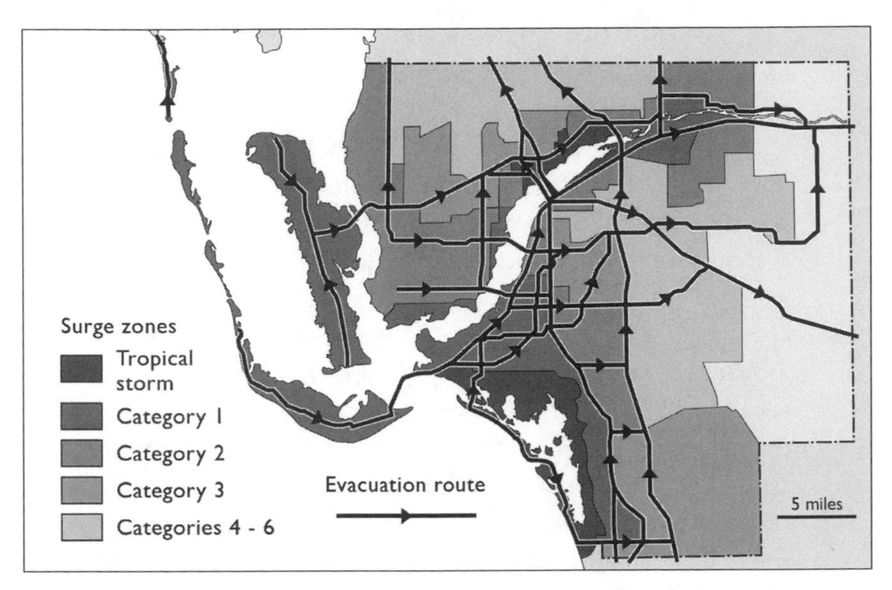

Figure 9.7. Surge-risk zones and evacuation routes for Lee County, Florida. Adapted from Lee County, Florida, Emergency Management, *Lee County Elevation and Evacuation Map.*

A third map (fig. 9.7) juxtaposes Lee County's hurricane evacuation routes with five levels of surge risk. Based on simulations for tropical storms as well as hurricane categories 1 through 6, the map relates surge heights to elevation above sea level and the local road network, which straightens out inundation boundaries.[41] In addition to providing rapid egress, highways and arterials are convenient landmarks, more easily communicated in emergency broadcasts than contour elevations. The map's five zones offer a selection of third shorelines, all simplified for human consumption—a valuable tool if users remember that a large, slow moving tropical storm with heavy rain but less intense winds can cause more flooding than a hurricane classified according to wind velocity.

Heavily dependent on historical data and computer simulations of coastal storms in some areas and tsunamis in others, the catastrophic flood line is essential for controlling development and planning evacuations.[42] Although multiple definitions and problematic measurements make its cartographic display elusive and uncertain, this complexity in no way diminishes the fact that the third coastline is a reality that citizens and their elected officials must understand and appreciate. In making the flood hazard visible, maps help protect life and property.

Public awareness is an awesome challenge for scientists who map risk for areas like coastal Louisiana, where local attitudes toward looming disaster have been fatalistic if not delusional. Well before summer 2005, residents and local officials as well as FEMA were well aware of the potential for catastrophic flooding. For decades surge models had described the devastating consequences of a direct hit on New Orleans, much of it well below flood level and separated from the Mississippi River and Lake Pontchartrain by fragile levees. In October 2001, for instance, a *Scientific American* article titled "Drowning New Orleans" reported that Louisiana State University scientists "who have modeled hundreds of possible storm tracks on advanced computers, predict that more than 100,000 people could die."[43] And in October 2004, less than a year before Hurricane Katrina, a *National Geographic* article titled "The Big Uneasy" described a scenario in which the city "was buried under a blanket of putrid sediment, a million people were homeless, and 50,000 were dead."[44] That the September 2005 storm proved less fatal—perhaps 2,000 dead but a half million homeless[45]—reflects Katrina's veering to the right as it approached the coast. A lucky break this time doesn't nullify forecasts of a far greater disaster.

TEN

RISING SEAS, ERODING SURGE

Few maps dare show it, but a fourth coastline reflecting rising sea level provocatively portrays the likely long-term impact of coastal erosion, climate change, and a land surface that subsides and rebounds as continents drift and glaciers retreat. Increasing concentrations of greenhouse gasses are steadily warming the atmosphere, which is slowly heating the oceans, which are predicted to lift sea level between eighteen and fifty-nine centimeters by 2100, largely because the world's seas hold massive amounts of water, which expands slightly when heated.[1] Although confident of global warming, scientists don't fully understand the complex interaction of atmospheric temperature, ocean currents, and polar ice shelfs. Iffy predictions of shifting currents suggest that glacial melting in Greenland and Antarctica could elevate sea level seven and sixty-five meters, respectively.[2] While rising seas seem less immediate a threat to coastal residents than storm surge and coastal erosion, maps of our fourth coastline are forcing policy makers in coastal states to pay attention to climate research.

Numerical forecasts of sea level rise are comparatively new. A century ago geographers used maps to infer that shorelines were advancing inland in some places and retreating seaward in others, but fragmentary tidal records and distracting hypotheses thwarted numerical prediction. The dominant theory was the "normal erosion cycle" derived from Charles Darwin's notion of organic evolution and promoted eloquently by William Morris Davis, a Harvard professor who imagined spasms of rapid continental uplift separating much longer periods of sustained erosion during which landscapes advanced through distinct stages of youth, maturity, and old age.[3] A cycle culminated in a low-relief surface called a peneplain (meaning "almost a plain"), which continued to erode until renewed uplift triggered another round. In 1919, Douglas Johnson, a protégé of Davis, extended his mentor's theory to coastal environments in *Shore Processes and Shoreline Development,* which included block diagrams differentiating "early youth" and "youth" stages for a hypothetical "shoreline of submergence" (fig. 10.1).[4] The principal agents were wave erosion and the riverine delivery of the sediment with which surf and longshore currents sculpted beaches, spits, and other coastal landforms. Johnson argued that "oscillations in the level of land or sea [could] leave a shoreline with a variety of features, some of which resulted from submergence, others from emergence," but his book reflected scant curiosity about the forces driving these oscillations and little interest in measuring their rates.[5]

Mapmakers were not so dismissive. The tide gauges that provided a vertical foundation for coastal charts and topographic maps registered noteworthy disparities in mean sea level from season to season, year to year, and place to place. While local weather and regional storm patterns accounted for some of this variation, data revealed erratically rising seas undermining the longevity of established topographic benchmarks. Particularly puzzling was an apparent increase from 1909 to 1919, following a decade of comparative stability. In 1925, U.S. Coast and Geodetic Survey tides expert Harry Marmer questioned conventional wisdom that attributed any increase in sea level to coastal subsidence, which at the time was believed to occur at a rate of two feet a century. "May not the mean level of the sea itself be changing, owing to a change in the volume of the ocean basins or to a change in the volume of water?" he asked in a *Geographical Review* article.[6] Although a rising sea and stationary coast can produce the same effect as a stable sea and subsiding coast, scientists impressed with Earth's capacity for mountain building considered the latter explanation far more plausible.

Recent evidence indicates that the East Coast has experienced

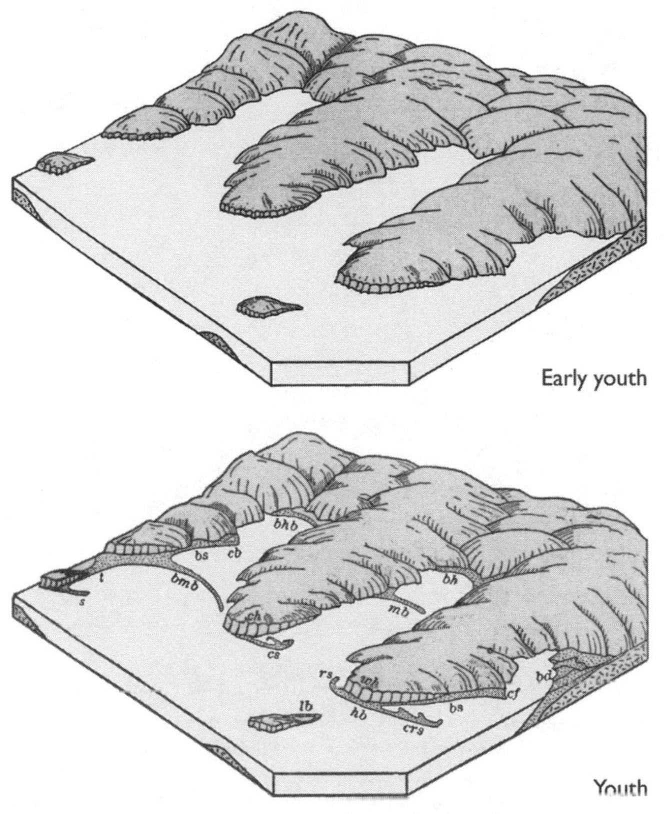

Figure 10.1. A crenulated shoreline characterizes the "early youth" (*above*) of an emerging shoreline in contrast to the beaches (*bh, bs, bs*), bars (*bmb, lb, mb*), and spits (*s, cs, crs*) that typify the coastline's more nuanced "youth." Modified from Johnson, *Shore Processes and Shoreline Development*, 275, 283.

rising seas as well as coastal subsidence at respective rates of eighteen and twelve centimeters over the last hundred years.[7] Although the resulting relative rise of thirty centimeters is slightly less than half the two-foot increase accepted uncritically a century ago, a likely acceleration of sea level rise in the current century is a distinct threat to low-lying areas like the Mississippi Delta, south Florida, and eastern North Carolina. Although a rise of one or two feet in a hundred years might seem inconsequential, the chain of barrier islands in North Carolina known as the Outer Banks is particularly vulnerable because elevated seas intensify the impact of storm surge and wave action.

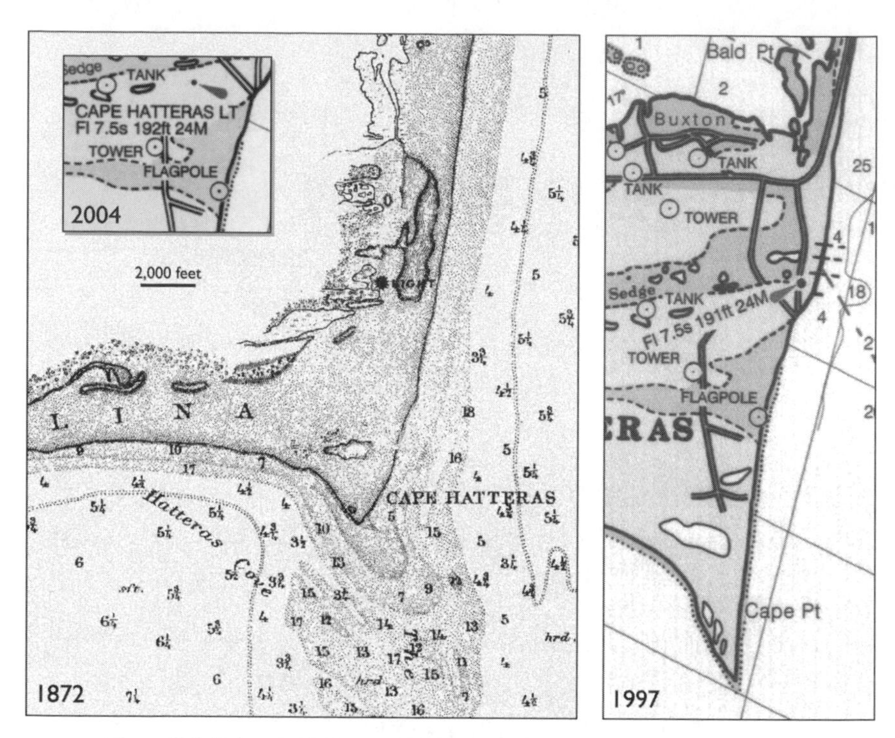

Figure 10.2. Between 1872 and 1997, the shoreline moved dangerously close to the Cape Hatteras lighthouse, which the National Park Service moved inland in 1999, as indicated by the 2004 inset at the upper left. From U.S. Coast Survey chart 416, Hatteras Shoals, 1:80,000 (1872), and NOAA nautical chart 11555, Cape Hatteras: Wimble Shoals to Oracoke Inlet, 1:80,000 (1997 and 1994 editions). On the more recent charts a pointer (printed in magenta) and dot indicate a single flashing light (Fl), 191 or 192 feet above the ground, with a period of 7.5 seconds.

These narrow islands were under assault well before climatologists began to worry about global warming, as is apparent in the saga of Cape Hatteras Lighthouse. Built in 1870 on a foundation 1,500 feet from the shoreline, the 200-foot structure was a mere 150 feet from the sea in 1999, when the National Park Service paid $10 million to move it 2,900 feet farther inland (fig. 10.2).[8] Because of an ominous acceleration in sea level rise suggested by data for the 1990s, this costly relocation is hardly an ultimate solution. Rising seas and development schemes that choke off the supply of sand could obliterate much of the Outer Banks by 2300.[9]

Close comparison of the chart excerpts in figure 10.2 reveal a marked alteration of the coastline between 1872 and 1997, when the Hatteras shoreline extended a mile south of its earlier position.[10] But there's more to the story. In 1936, with the sea only 100 feet from the lighthouse, the Coast Guard moved its flashing beacon to a steel-frame tower farther inland. To prevent storm surge from reaching the historic structure, the Civilian Conservation Corps built an artificial dune in the late 1930s. These efforts worked for a while, but sandbags had to be added around the foundation, and additional sand was pumped in at various times in a costly short-term fix known as beach nourishment. The Coast Guard reactivated the lighthouse in 1950, and transferred ownership to the National Park Service. As a further defense against erosion, the Navy built three groins in 1969—and rebuild them six years later, after a storm wiped them out, along with much of the beach. (Groins are rigid structures extending perpendicularly from the shoreline to trap sand and reduce beach erosion. Thick black dashed lines on the 1997 chart represent the three groins, northeast of the lighthouse and typically covered at high tide.) Reconstructed groins and new sand could not prevent storm surge from advancing within seventy feet of the foundation during a March 1980 storm. Well aware by 1987 that the Outer Banks were moving slowly toward the mainland, the National Park Service sought the advice of the National Research Council, which reviewed the data and recommended relocation.[11] By then, most earth scientists recognized human-induced climate change as a key factor in sea level rise.

Among the earliest cartographic efforts to dramatize rising seas was a 1980 map showing the Potomac River encroaching on Capitol Hill. Writing in the *Annual Review of Energy*, climatologists Stephen Schneider and Bob Chen speculated on the impact of increasing concentrations of carbon dioxide, which some scientists feared might warm the atmosphere by as much as 5°C by the year 2000. Among other consequences, massive melting of the West Antarctic ice shelf could elevate water level worldwide by five meters in a few decades. While wary that "the possible rise in sea level is certainly not the most immediate, and perhaps not even the most important potential CO_2–induced environmental effect,"[12] Schneider and Chen embellished the U.S. Geological Survey's Washington West 7.5-minute topographic quadrangle map with thick 15- and 25-foot contours representing shorelines 4.6 and 7.6 meters above mean sea level.[13] With much of the former swamp underwater, they observed, "one could launch a boat from the west steps of the United States Capitol . . . and row to the White House South Lawn."[14] In case Congress and the president didn't get the message, they plotted similar flood lines on hundreds of other topo quads along the country's Atlantic, Gulf, and Pacific coasts, and estimated that 15- and 25-foot inundations would displace 5.7 and 7.8 million people, respectively.

A potentially significant exemplar of persuasive cartography, their depiction of the Washington Monument sitting in a vastly enlarged Tidal Basin fared poorly as a two-page black-and-white journal illustration. Bad photography or sloppy photoengraving lost many of the topographic map's key details, originally in color, and the gutter between the pages swallowed symbols and distorted distances. Because the hypothetical shorelines are worth savoring, especially along the Mall between the Capitol and the White House, I scanned a portion of the journal page, traced its bold contours in Freehand, and overlaid the result on the corresponding section of a more recent USGS topographic map, painstaking edited in Photoshop. My reconstructed contours in figure 10.3 capture the threat nicely. Dashed lines—I could match the thickness of the original but not its individual dashes—demonstrate how a rise of fifteen feet might bisect the Ellipse across from the White House, swallow the Reflecting Pool just west of the Capitol, and inundate the National Gallery of Art and the complex of office buildings between Constitution and Pennsylvania Avenues known as Federal Triangle. The solid line representing a twenty-five-foot rise would invade the south lawn of the White House, establish a beachhead in front of the Capitol, and encircle a new island dominated by the Department of Agriculture monolith on Independence Avenue. The Smithsonian Institution's imposing main building, known as the Castle, sits above the twenty-five-foot line, but most of its constituent museums would be swamped.

While Schneider and Chen might have been the first atmospheric scientists to present a detailed, highly localized depiction of a plausible (if not immediately imminent) inundation, the last page of their article points to a largely unknown cartographic warning issued nine years earlier. On the final page of their article, beneath the last

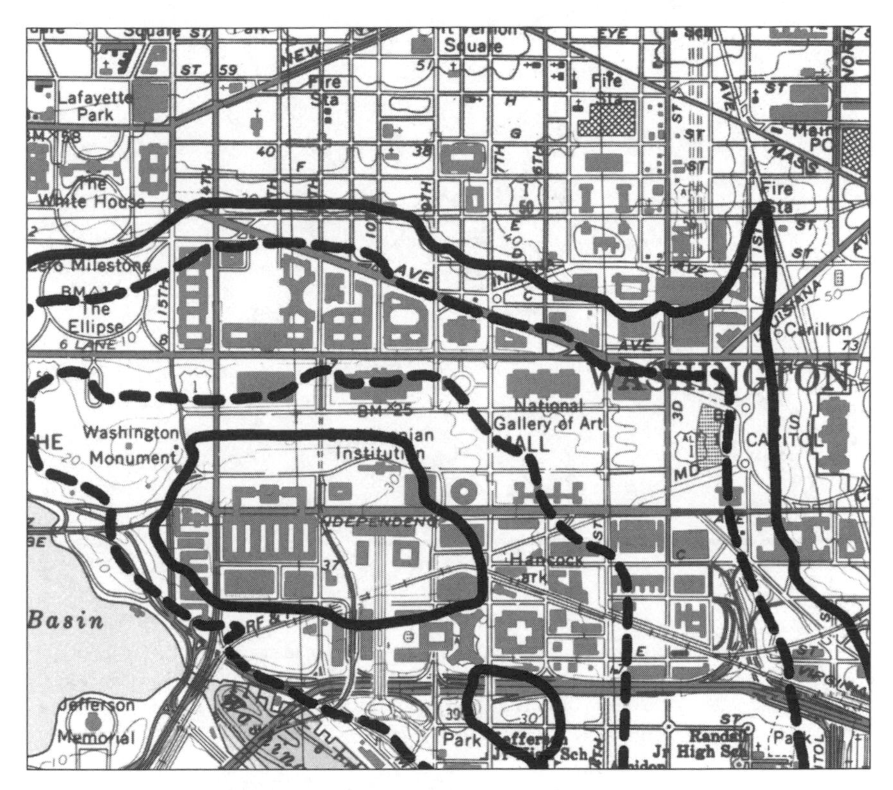

Figure 10.3. A reconstruction of part of the inundation map of Washington, D.C., published in 1980 by Stephen Schneider and Bob Chen. The dashed and solid lines represent elevations of 15 and 25 feet above mean sea level.

of their 131 bibliographic citations, is a "NOTE ADDED IN PROOF" reporting that "the authors have become aware of a study which examines a rise of 100m due to a melting of all glacial ice (131)."[15] Authored by University of North Carolina geographer Richard Kopec, the study appeared in the *Journal of Geography*, published mostly for secondary school teachers and apparently not on the radar of climate change scientists, for whom a 100-meter rise far exceeded the most dismal scenarios. "But let's be academic," mused Kopec, and "accept the premise that due to overpopulation and uncontrolled abuse of the atmosphere, mankind will cause the climate of the earth to change and this change will result in rising air temperatures . . . sufficient to melt all remaining ice forms, releasing their stored water to the oceans."[16] To describe the impact of a 100-meter (330-foot) sea

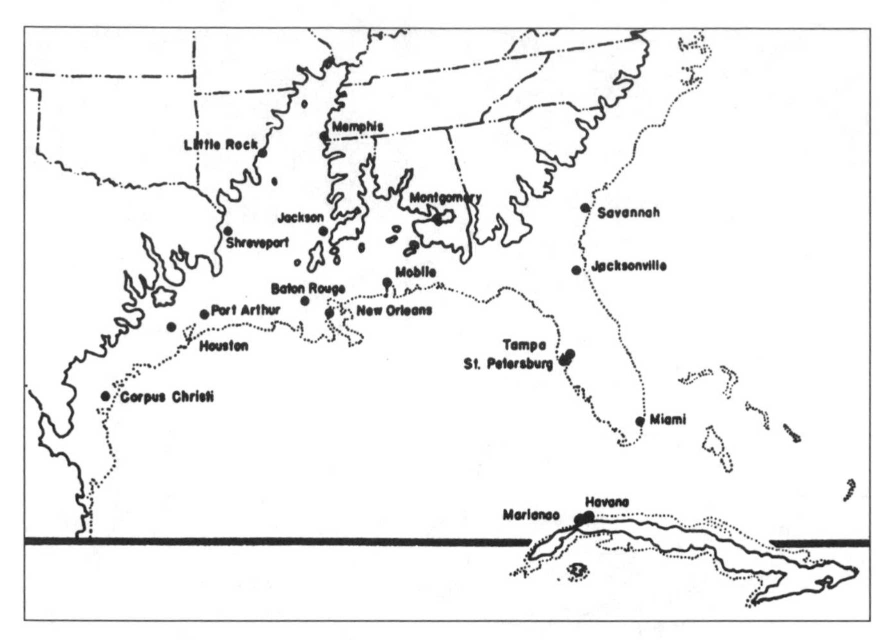

Figure 10.4. Southeast portion of Richard Kopec's 1971 map describing the impact of a 100-meter sea level rise on the United States. Thick black line is the original map frame, breached by inclusion of Cuba. From Kopec, "Global Climate Change and the Impact of a Maximum Sea Level on Coastal Settlement," 548.

level rise, deemed "remote but not impossible" by 2050, Kopec portrayed the coastline's current and forecast positions with dotted and solid lines, respectively, and identified newly inundated major cities by name.[17] A world map foretells invasive embayments in Brazil and Australia, massive flooding in Bangladesh, and an amalgamation of the Mediterranean, Black, and Caspian Seas. A separate map depicts an equally gloomy future for the United States, where rising seas would cover much of the Southeast, including the entire state of Florida (fig 10.4). A mildly subversive southerly extension of Kopec's map shows comparatively mountainous Cuba as mostly high and dry—a potential haven for boat people fleeing Florida?

Mapping the projected consequences of sea level rise would be easy had the unduly pessimistic estimates of the 1970s fostered an uncritical use of existing maps. Detailed topographic maps typically have a contour interval of 10 feet, which yields hypothetical shorelines at 10 feet, 20 feet, 30 feet, and so on. To plot the extent of a 15-

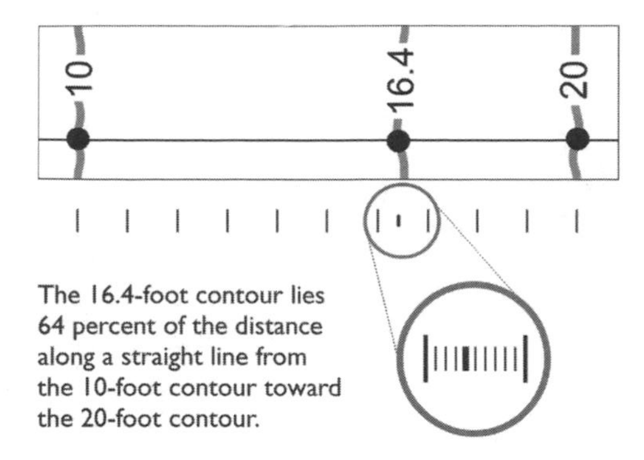

Figure 10.5. Interpolation of a dot on the 16.4-foot (5-meter) contour line between an existing map's 10- and 20-foot contours. To plot the 5-meter contour, a mapmaker would link up similar dots along other transects between the two contours.

foot rise, Schneider and Chen conveniently connected a moderately dense series of dots midway between the 10- and 20-foot contours. Fourth shorelines at other levels can be interpolated if one presumes a uniform slope between adjacent contour lines. Figure 10.5 shows how to position a dot on the contour at 16.4 feet, which equals 5 meters, by dividing a straight line between points on the 10- and 20-foot contours into ten equal parts, subdividing the seventh interval (between 16 and 17 feet) into ten smaller parts, and aligning the dot with the fourth intermediate tic-mark.

Unfortunately for map analysts, when climatologists and coastal scientists reached a consensus that sea level rise was real, they foresaw inundations of generally less than a meter (3.3 feet) by 2100 and not quite two meters (6.6 feet) by 2200.[18] I say "unfortunately" because straight-line interpolation between elevations of 0 and 10 feet ignores coastal wetlands, which typically produce an elbow-shaped profile like that in the right half of figure 10.6. Along a marshy shoreline, linear interpolation is justifiable only after the lower portion of the slope is pushed away from the shoreline toward the dry edge of the wetland. Otherwise, the map underestimates the impact of rising seas.[19] Although topographic maps for gently sloping coastal areas typically have 5-foot contours, linear interpolation is problematic for scientists concerned with the smaller rise predicted by the Intergovernmental Panel on Climate Change (IPCC).

Inundation is not the only consequence of sea level rise. An elevated sea carries storm surge farther inland, exposes new land to wave action, and disturbs an otherwise stable beach slope. In the early

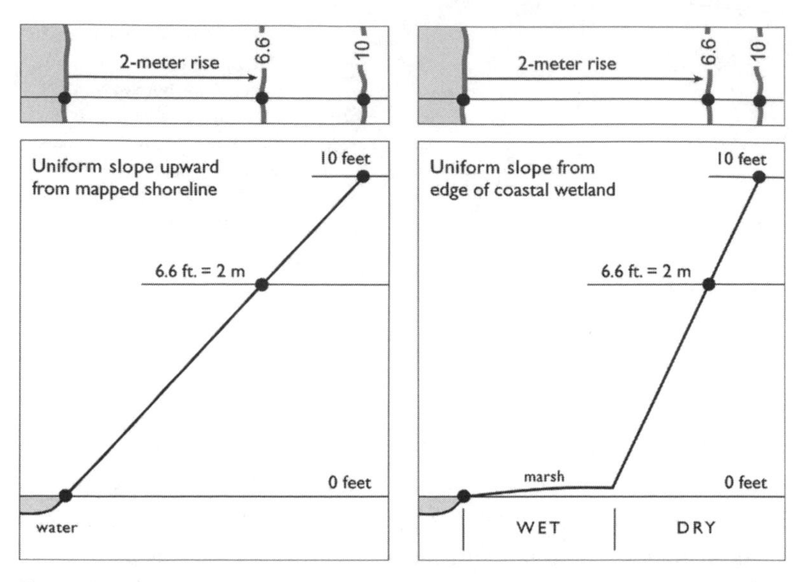

Figure 10.6. Unless marshland is considered (*right*), the straight-line interpolation (*left*) of the two-meter contour will underestimate vulnerability to sea level rise.

1960s, Danish-born civil engineer Per Bruun derived a mathematical relationship describing this effect on sandy beaches, which erode easily. According to Bruun, beaches adjust to rising water by forming a new profile. As figure 10.7 illustrates, a stable profile requires a transfer of sand from the beach to the offshore bottom, which rises in response to sea level rise.[20] The resulting displacement of sand moves the shoreline farther inland, well beyond the point of mere inundation. Although the "Bruun Rule" ignores horizontal movement of sand along the shore and works poorly near inlets through barrier islands, coastal engineers use it to project shorelines twenty-five years or more into the future.[21] Other forecasts rely on rates of retreat calculated from historic shorelines on Coast and Geodetic Survey T-sheets (fig. 2.5).[22] Even so, multiple sources of uncertainty as well as the effects of coastal engineering elsewhere along a beach undermine the reliability of these forecasts, which are often used to argue for beach nourishment and similar strategies for "saving" eroding shorelines.

Coastal engineering might halt shoreline migration temporarily, but not without consequences. Forecasts are particularly grim where tidelands regulations ban construction near the water but allow dikes

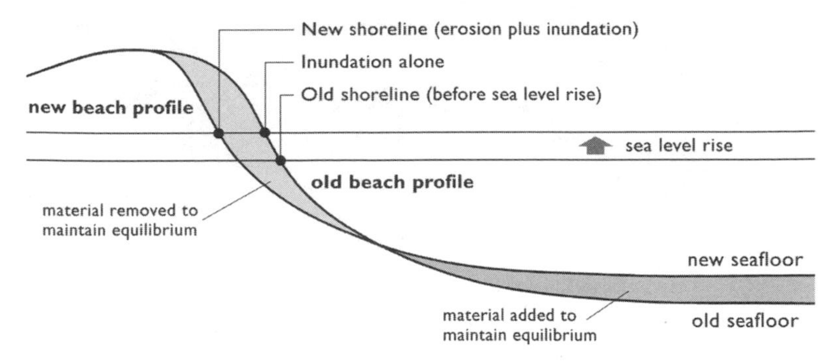

Figure 10.7. According to the Bruun Rule, beach profiles reestablish equilibrium after sea level rise by eroding landward past the inundation level.

and revetments to protect new roads and buildings set well back from current shoreline.[23] (A dike is a seawall, usually erected above the current high-water line, whereas a revetment resists erosion by covering, or "armoring," a sandy beach with large rocks.) Although developers and wealthy retirees might think the seawalls prudent, an accelerated sea level rise will eventually expose these defenses to wave erosion, which obliterates the beach and leads to still higher seawalls or a more heavily armored shoreline, no longer buffered by beaches and salt marshes. Jim Titus, who manages the Sea Level Rise Project at the U.S. Environmental Protection Agency, warns that short-sighted policies will rob future generations of sandy beaches. We need to get out of the way of rising seas as they migrate inland, he argues, or spend millions of dollars yearly on beach nourishment. His solution is "rolling easements," a strategy that allows development along the shoreline but bans seawalls and other obstructions.[24]

Titus believes maps can help public officials understand impacts and options. Because detailed coastal-zone mapping is a slow and complex process, constrained by the accuracy of elevation data and other local information, he began with a broad assessment of vulnerability to sea level rise along the Atlantic and Gulf coasts. His first maps, released in 2000, show where the threat is most worrisome at the state and regional levels. Compiled from admittedly coarse but readily available computer data—NOAA provided the shorelines and the U.S. Geological Survey contributed inland elevations—the maps use a 1.5-meter (4.9-foot) contour line to identify low-lying areas par-

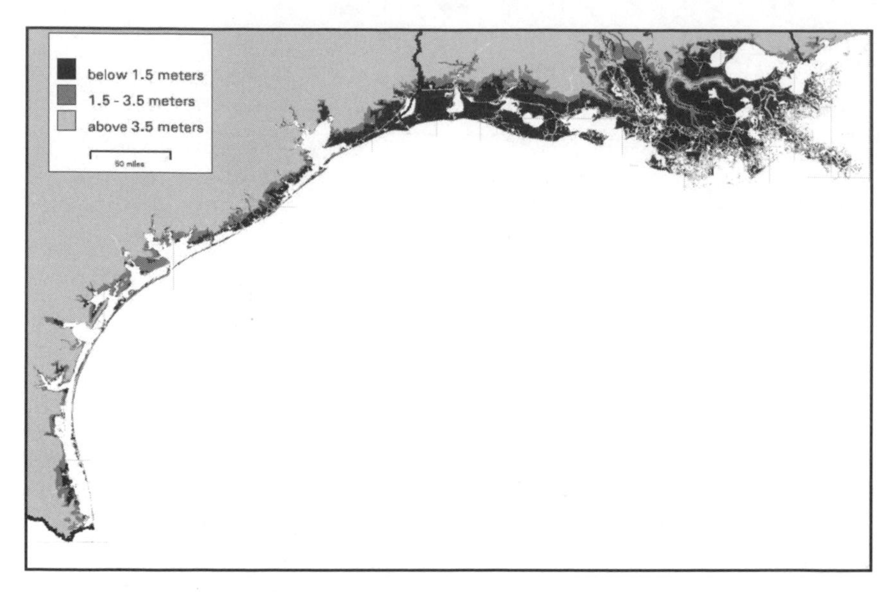

Figure 10.8. Lands Vulnerable to Sea Level Rise: West Gulf Coast. The color version of this map depicts the black areas (under 1.5 meters) in red and the dark gray areas (between 1.5 and 3.5 meters) in bright blue. From the U.S. Environmental Protection Agency, Global Warming Publications, http://yosemite.epa.gov/OAR/globalwarming.nsf/UniqueKeyLookup/SHSU5BUN3P/$File/gulfwestbw.pdf.

ticularly susceptible to flooding within a hundred years (mapped in red) and a 3.5-meter (11.5-foot) contour to delineate areas "that might be flooded over a period of several centuries" (mapped in blue).[25] Published on the Internet at three different scales, Titus's elevation maps reveal vast areas with an uncertain future in south Florida and eastern North Carolina as well as along the eastern shore of the Chesapeake Bay and the Gulf Coast between Galveston and the Louisiana-Mississippi border.[26] His comparatively detailed map of the West Gulf Coast (fig. 10.8) underscores the ominous flood hazard in coastal Louisiana, where land subsidence related to extraction of oil and natural gas reinforces the sea level rise caused by climate change. Not a promising scenario for politicians eager to rebuild New Orleans.

Well aware that small-scale maps are of limited use as planning tools, the EPA began compiling large-scale coastal-zone elevation maps based on more accurate topographic information, including wetlands data, so that interpolated contours might more accurately reflect tidal marshlands (see fig. 10.6).[27] As spatially detailed as the

Geological Survey's large-scale, 1:24,000 topographic sheets, these new maps focus on elevations below six meters and rely on color symbols to differentiate categories at one-meter intervals. Light blue identifies land below 0 meters as tidal wetland, while dark green, light green, yellow, orange, red, and brown signify successively higher (and progressively less threatened) elevations. Jim Titus let me look at some of the maps if I promised not show or publish them before the EPA authorized their release—politically sensitive materials like these require a careful review by scientists and knowledgeable local officials.

More intriguing are the county-level planning maps designed to evaluate what Titus calls the "business as usual" response to sea level rise: elevating structures where feasible; defending infrastructure, resort hotels, and expensive homes with revetments, sea walls, or beach nourishment; and letting the eroding shoreline retreat unimpeded elsewhere.[28] Because the impact of rising seas depends on the response of property owners as well as state regulations and local codes, the new EPA maps use contrasting colors to distinguish low-elevation locations according to whether artificial shoreline protection is almost certain (brown), likely (red), unlikely (blue), or not allowed (light green).[29] The brown areas include coastal cities and resorts like Atlantic City, New Jersey, and Ocean City, Maryland, where property owners and local officials will vigorously resist rising waters, as well as the moderately developed shorelines of relatively calm bays, where protective measures are effective and less costly. Coastal engineering seems likely but less certain in the red areas for diverse reasons, including conservation easements, bans on federal subsidies, or a lack of sewer and water systems. Red's customary association with danger makes a plea for informed decision making. By contrast, blue identifies areas where farming, conservation easements, or strict regulations make shoreline armoring "unlikely," while light green highlights publicly and privately owned conservation land, where existing laws or carefully worded covenants preclude dikes and seawalls.

Compiled with the assistance of state and county planning officials, these maps were still under review. The examples I was allowed to view include two- and five-meter elevation contours, which Titus

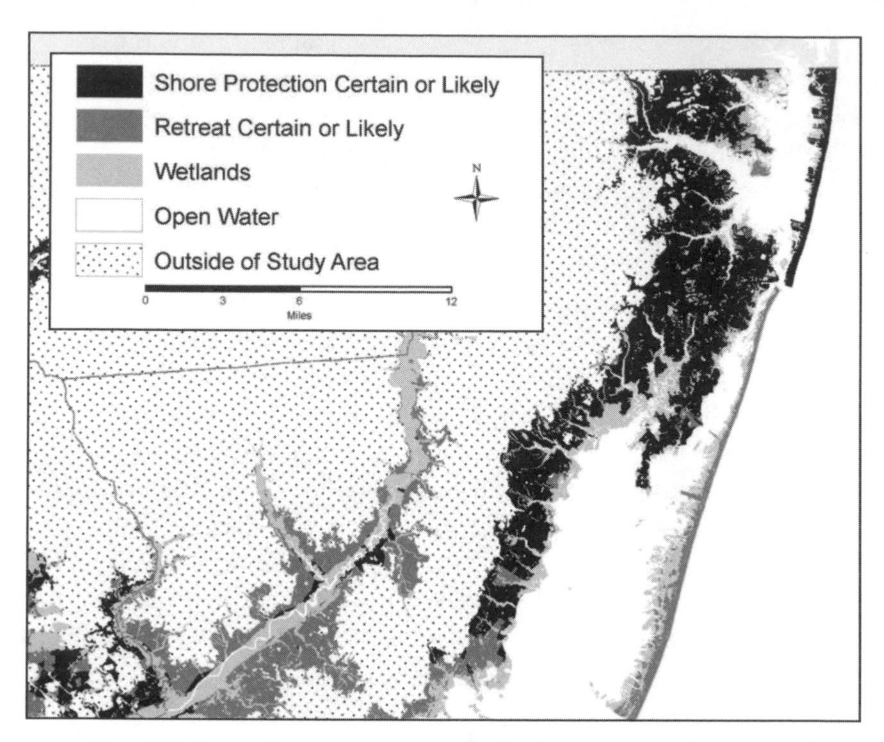

Figure 10.9. Excerpt from a page-size map showing "business as usual" approaches to shore protection in Worcester County, Maryland. This black-and-white version shows brown and red areas on the color map in black, and blue and light green areas in dark gray. From Titus, "Maps That Depict the Business-As-Usual Response to Sea Level Rise in the Decentralized United States of America," 19. The two barrier islands, Fenwick and Assateague, are separated at the inlet just south of Ocean City, located on Fenwick Island, where high property values make shoreline protection inevitable.

used to delineate two levels of urgency. For each color-coded response category, a brilliant, full-strength hue underscores the more immediate threat to elevations below two meters, while a lighter, desaturated color connotes the less imminent danger to elevations between two and five meters.

Black-and-white versions must combine categories. The example in figure 10.9, unveiled in Paris in late 2004 at the UN-sponsored Global Forum on Sustainable Development, highlights radically different approaches for built-up and undeveloped parts of Worcester County, Maryland, on the Atlantic between Delaware and Virginia. Contrasting graytones indicate shore protection is either "certain or

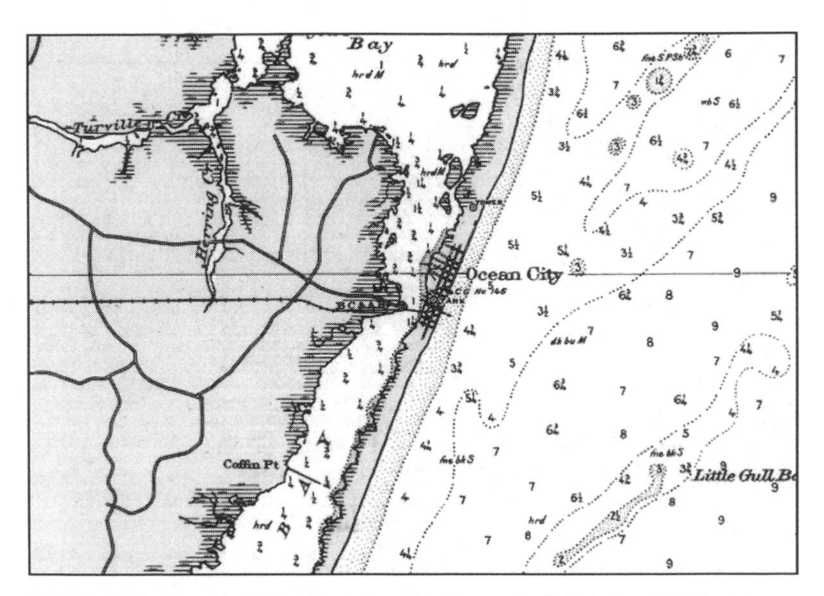

Figure 10.10. Portion of U.S. Coast and Geodetic Survey chart 1220, Fenwick Island to Chincoteague Inlet, published at 1:80,000 in 1920. Compare the continuous barrier island through Ocean City with the separate islands on the present-day map in figure 10.9.

likely" almost everywhere on the highly developed barrier island in the northeast part of the county known as Ocean City, whereas retreat is "certain or likely" immediately south, where the barrier island is offset noticeably to the west. The significance of this offset is apparent on a 1920 Coast and Geodetic Survey nautical chart that shows a continuous barrier island through Ocean City (fig. 10.10). In 1933 a devastating hurricane opened the inlet and severed the resort from the southern section, now shared by Assateague Island National Seashore and a state park.[30] Maintained for boat-owners, the inlet became a dividing line between protected and unprotected shorelines. While tax dollars help Ocean City defend its high-rise condos and renovate its beaches, Assateague Island retreats westward and keeps its natural seashore.

New worries about accelerated melting in Greenland and Antarctica make Titus's two- and five-meter contours look benignly optimistic.[31] What's uncertain is the combined effect of slightly warmer air atop the ice sheet, which rests on land while moving very slowly toward the sea, and slightly warmer water beneath the ice shelf,

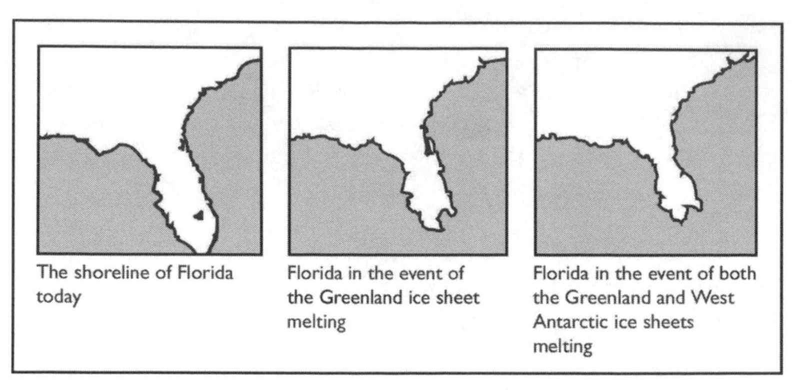

Figure 10.11. Byron Parizek's maps describing the shrinkage of Florida's coastline following the melting of the Greenland and West Antarctic ice sheets. From Bellassen and Chameides, *High Water Blues,* 2005 Update, 10.

which floats on the ocean and is vulnerable to increased melting from below. Surface melt trickling downward through fissures reduces friction at the bottom of the ice sheet, which can slide more easily toward the sea. Melting along the coast contributes to this accelerated movement by reducing the damming effect of a stable ice shelf, which helps keep the ice sheet up on the land and out of the sea. Computer simulations suggest that "basal sliding" and related effects might add 15 to 40 centimeters to the Greenland ice sheet's projected contribution of 55 to 360 centimeters to world sea level rise by 2500.[32] Antarctica's contribution would push the fourth coastline even higher. Byron Parizek, a glaciologist who uses computer models to probe relationships between climate change and sea level, relies on maps to dramatize the potential impact on features like Florida's coastline (fig. 10.11). Presented at congressional briefings and public lectures to "drive home the importance of understanding ice dynamics in a warming world," his maps underscore the threat to world peace and national security of abrupt and unexpected climate change powerful enough to trigger mass migrations.[33]

ELEVEN

CLOSE-UPS AND COMPLEXITY

I round out this exploration of the evolution, uses, and impacts of shoreline delineations with a survey of cartographic ingenuity in portraying land-water boundaries. The examples that follow reflect not only the numerous maps and charts encountered while researching this book but also my careful scrutiny—bordering on suspicious behavior in at least one security guard's opinion—of an impressive exhibit called "Treasured Maps" at the New York Public Library.[1] Instead of trying to ferret out famous firsts and quirky curiosities, I focus on standard solutions, which collectively illustrate the mapmaker's traditional appreciation of symbols that are readily recognized and reliably decoded.

Among the oldest solutions is a simple shoreline embellished with perpendicular place names. The detail in figure 11.1, enlarged from a 1582 portolan chart by Spanish cartographer Joan Martines, is typical of the sailing charts that emerged in the late thirteenth century.[2] Intended to illustrate verbal sailing directions published in pilot books known as *portolani,* these maps are dominated by a network of straight-line sailing routes called rhumb lines, which occasionally intersect at prominent compass roses. Labeling of ports and fishing villages on the landward side of the coastline minimizes visual conflict with the rhumb lines. Town icons like those in the excerpt sometimes look upside-down because users were expected to rotate the chart. A common navigation tool shortly after European mariners

Figure 11.1. Landward place names on Joan Martines's 1582 portolan chart of the eastern Mediterranean depict the shoreline as a boundary between inhabited places and navigable waters. From Stevenson, *Portolan Charts: Their Origin and Characteristics with a Descriptive List of Those Belonging to the Hispanic Society of America*, opp. 58.

mastered the magnetic compass, the medieval portolan chart anticipated Gerard Mercator's innovative 1569 world map, on which all angles are correct and any straight line represents a constant direction.

Renowned for the map projection that carries his name as well as the pioneering world atlas that established *atlas* as the generic term for a book of reference maps, Mercator recognized the value of using map symbols to differentiate known and comparatively unknown shorelines, as shown in figure 11.2, an enlarged portion of his less famous 1538 world map. Because the South Pole lies toward the upper left corner of this excerpt, the tip of South America is at the bottom, the Antarctic Peninsula is at the top, and the Atlantic and Pacific Oceans lie to the left and right, respectively. Short, thin hatch lines, characteristically horizontal and closely spaced, extend seaward from a thin, crisp, continuous shoreline marking the more reliably explored Atlantic coasts. Horizontal hatching is a common

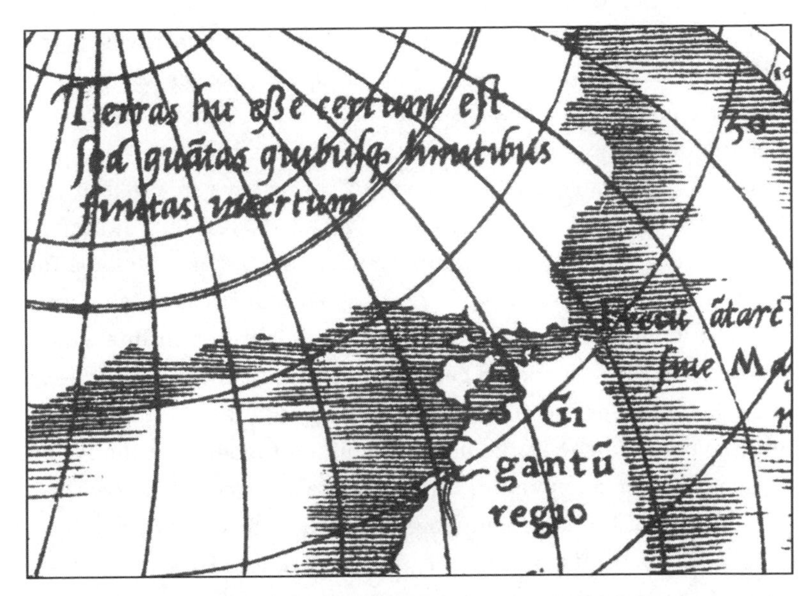

Figure 11.2. Gerard Mercator's 1538 world map uses subtly different shoreline symbols to differentiate known and unexplored coastlines. From Nordenskiöld, *Facsimile Atlas to the Early History of Cartography with Reproductions of the Most Important Maps Printed in the XV and XVI Centuries*, plate XLIII.

feature of world and regional maps engraved on malleable copper printing plates, which began to replace wood engraving in the late fifteenth century.[3] Particularly distinctive is the thin shoreline's subtle omission from the largely unexplored Pacific and Antarctic coasts. Mercator's portrayal of questionable coastlines with unanchored hatching impressed biographer Nicholas Crane, who saw the map as "a progress report on exploration."[4]

Unlike Mercator, who catered to well-educated and generally affluent clients, chart makers like Joan Martines served working mariners, who required a variety of purposeful coastline symbols. A close look at Martines's portolan chart (fig. 11.1) reveals a carefully coded treatment of what is now the northeastern coast of Egypt and the maritime margins of Israel and Gaza. Rounded indentations identify small bays or otherwise sheltered anchorages separated by blunt, exaggerated headlands, while short interruptions in the thick, dark shoreline mark the entrances of rivers, large and small. The northward-flowing Nile and its delta are easily identified. A chain of tiny dots running eastward from the delta and then northward represents shoal (shallow) water and other hazards, larger dots signify small islands, and tiny crosses mark dangerous offshore rocks.[5]

Thick, visually prominent cartographic coastlines were no less crucial on the systematically surveyed charts that emerged in the nineteenth century, according to pioneer U.S. Coast Survey topographer Henry Whiting. In explaining nautical charting symbols for the agency's 1860 annual report, he argued that because "the limit of land and water is the most striking and important outline which exists in nature . . . it should be the strongest and most conspicuous line of boundary between natural features [on] a map."[6] What's more, Whiting opined, "the immediately adjoining topography should be distinctly and strongly represented [to] give a more striking contrast to the limits of the land and water [as well as to show the] outline of bays, coves, points, islands, &c., in strong relief." Equivalent treatment of the low-water shoreline and adjacent details was less important because hydrographers "could not develop all the irregularities of low water by the ordinary process of field-work," and because "the shore is more cautiously approached by vessels, and therefore less dangerous in navigation."[7] Safe sailing required hydrographers to focus on charting shoals, mudflats, rocks, and other hazards lurking offshore.

Today's mariners demand a detailed treatment of tidal influences on coastal hazards. Although water level can fluctuate widely over the lunar month as well as seasonally, modern nautical charts rely on established high-water and low-water chart datums to report whether a hazard is always, sporadically, or never visible at the surface.[8] As the corresponding plan and profile views in figure 11.3 illustrate, a rock bare at high water is outlined on detailed, large-scale charts by the same prominent, continuous black line used for the high-water shoreline. The buff color symbolizing dry land fills the closed curve, and a number reporting elevation above the high-water datum warns of possible inundation during an exceptionally high tide. By contrast, a tiny asterisk or irregular black line indicates a rock that "covers and uncovers" between high and low tide, while a cross marks a sunken rock, always submerged at low water. Hydrographers are especially wary of tall, sharp "pinnacle rocks,"

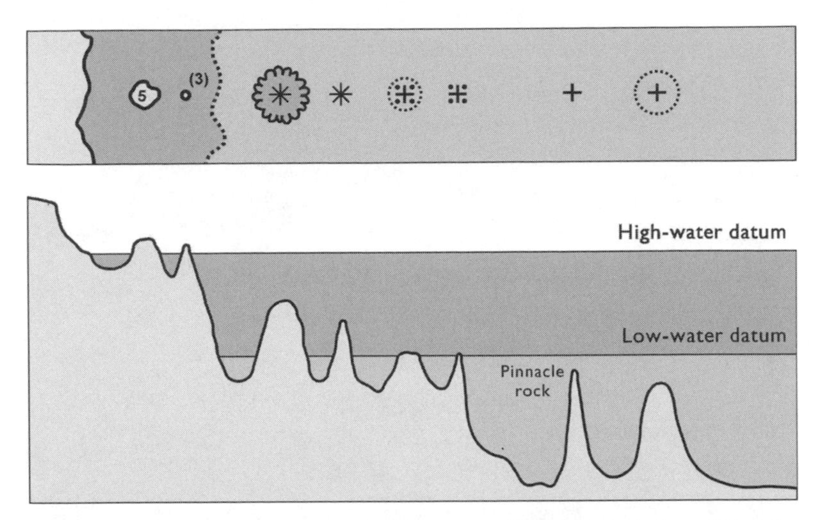

Figure 11.3. Schematic map (*above*) and profile diagram (*below*) illustrate the relationship of rock and islet symbols to the high-water and low-water chart datums. Asterisk and cross symbols are enlarged for clarity.

extending upward from a safely deep seabed. A "rock awash," visible if not fully uncovered at low tide, receives the special treatment of a cross with tiny dots in its corners. Often the asterisk, cross, or cross-with-dots appears within a dotted circle or an irregular dotted line indicating the hazard's extent.

As shown on large-scale nautical charts, the coastline is more than just a line. NOAA's Office of Coast Survey, which explains the layout, coding, and use of its navigation charts in a ninety-five-page booklet titled *Chart No. 1*, needs four pages to describe standardized symbols for various kinds of rocks, wrecks, and other offshore obstructions.[9] Features warranting special treatment include reefs, suspicious areas of discolored water, pilings or stumps, duck blinds (above water or submerged), fish traps, and oyster beds. As the examples in figure 11.4 illustrate, abbreviations obviate an otherwise confusing assortment of abstract and pictorial symbols.

Color tints highlighting shallow water reinforce tide-based coding of rocks and other hazards. Typically light olive-green fills the area between a contemporary chart's high- and low-water shorelines, while light blue accents shoal water, the depth of which

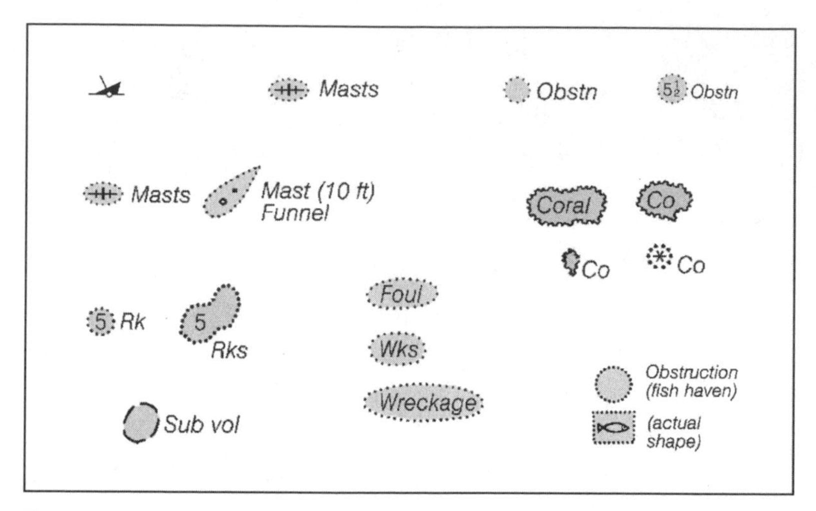

Figure 11.4. Examples of hazard symbols and abbreviations on NOAA's large-scale nautical charts. From National Oceanic and Atmospheric Administration, *Chart No. 1,* 43–46.

depends on scale—a small-scale chart might place the shoal line at 18 feet while a large-scale chart treats only depths less than 6 feet as shallow water. Some charts enhance their depth contours with multiple bands of blue. On the printed 1:80,000 chart from which I extracted a snippet for Five Islands, Maine (*left panel* of fig. 6.7), a near-shore band of light blue underscores depths less than 18 feet while a zone of noticeably lighter blue highlights depths between 18 and 30 feet. On the corresponding electronic chart (fig. 6.7, *right*), depth contours at 3.6, 5.4, and 18.2 meters (12, 18, and 60 feet) bound progressively lighter bands of blue, while beyond the 18.2-meter line a fainter blue-gray representing deeper water extends seaward indefinitely. Some commercial British charts reverse the scheme, with darker blues for deeper water.[10]

Soundings and annotations describing navigation aids in coastal waters further accentuate the cartographic coastline as a magnet for map-worthy features. On the landward side of the line, the modern nautical chart emphasizes beacons, church spires, and other landmarks by omitting features irrelevant to mariners. Although streets and prominent structures near docks and anchorages are shown for

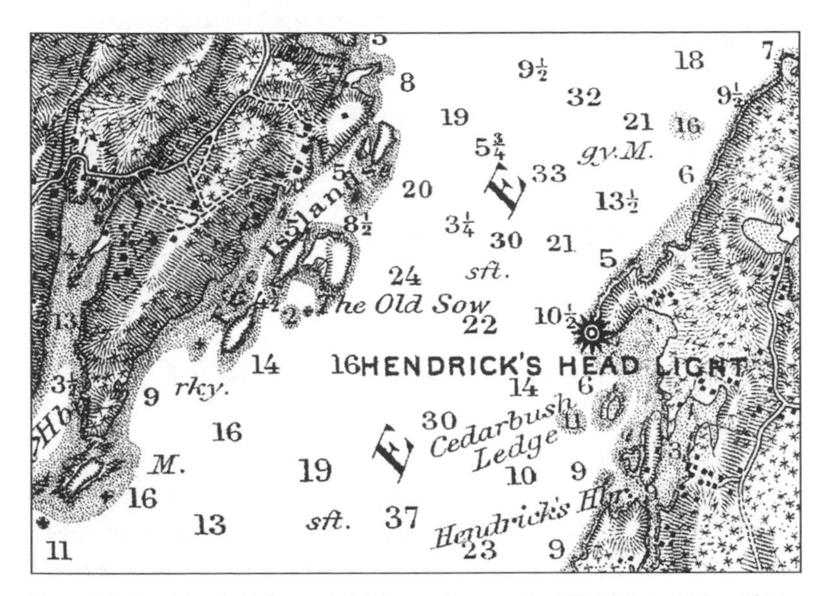

Figure 11.5. Five Islands, Maine, and vicinity, as shown on the 1:80,000 chart "Coast from Pemaquid Point to Seguin Island, Maine," bound facing p. 372 in U.S. Coast and Geodetic Survey, *Atlantic Local Coast Pilot, Sub-Division 4: White Head Island to Cape Small Point* (1879).

the convenience of sailors, chart makers purposely suppress details likely to create visual clutter. Anyone requiring a more comprehensive treatment of onshore territory should consult a U.S. Geological Survey topographic map, equally stingy with its offshore details.

This onshore-offshore division of labor is less apparent on some late-nineteenth-century American navigation charts. Established in 1879, the Geological Survey took several years to set up a map division and a few decades more to complete quadrangle maps for the nation's coasts, mapped earlier but less thoroughly by Coast and Geodetic Survey topographers. With survey results at hand and a clear need for greater interior detail, coastal cartographers supplemented the government's Coast Pilots with elegant relief maps for terrain within a half-mile to a mile of the shoreline (see pp. 84–85). Figure 11.5, enlarged from a chart included with the 1879 *Local Coast Pilot* for the stretch of coastline that includes Five Islands, Maine, describes landforms and local roads as well as nearby shoals and navigable waters. Hills and valleys are depicted by short strokes called hachures, which point upslope and suggest a three-dimensional relief model viewed from above. Absence of hachures signifies relatively level land, black rectangles represent houses and other buildings, and a generous sprinkling of tiny six-pointed asterisks indicates woodland. Labels identify features named in the text, while bold type and distinctive symbols mark important navigation aids like Hendrick's Head Light. A moderately dense pattern of tiny dots called stippling marks the extent of shallow water, within which small crosses indicate noteworthy hazards, described more fully in the Coast Pilot.

Disappearance of hachures from Coast and Geodetic Survey charts in the late nineteenth century reflects improved data-collection and reproduction techniques.[11] Although most changes were less dramatic, Coast and Geodetic Survey chart makers overhauled their graphic style several times between 1840 and 1960.[12] Particularly noteworthy is an emerging preference for a less cluttered map, exemplified by figure 11.6, which compares 1907 and 1927 editions of the 1:80,000 chart for the San Juan Islands, in northwest Washington. Fewer, more significant soundings and abandonment of the visually busy symbol for woodland allow a stronger focus on the coastline.[13] This more open design proved beneficial in later decades, when chart makers needed room for newly discovered submerged dangers; a growing list of buoys, beacons, channel markers, and military restrictions; and the special reference lines and annotations used with electronic position-fixing systems like Decca and Loran-C.

The 1927 excerpt (fig. 11.6, *lower*) reflects the concise graphic coding of Coast and Geodetic Survey charts engraved for black-and-white printing. Find the "white cliffs" in the upper right quarter of the example, and note the short, thin strokes on the inland side of the thick black high-water line. Consistently perpendicular to the coast, these closely spaced strokes indicate a steep, low bluff. By contrast, the longer strokes a quarter inch below the spot elevation 292 on the west side of the channel signify a substantially higher bluff. For another example, find "Cattle Pt.," examine the thick, high-water line extending westward to the edge of the map, and note the dotted

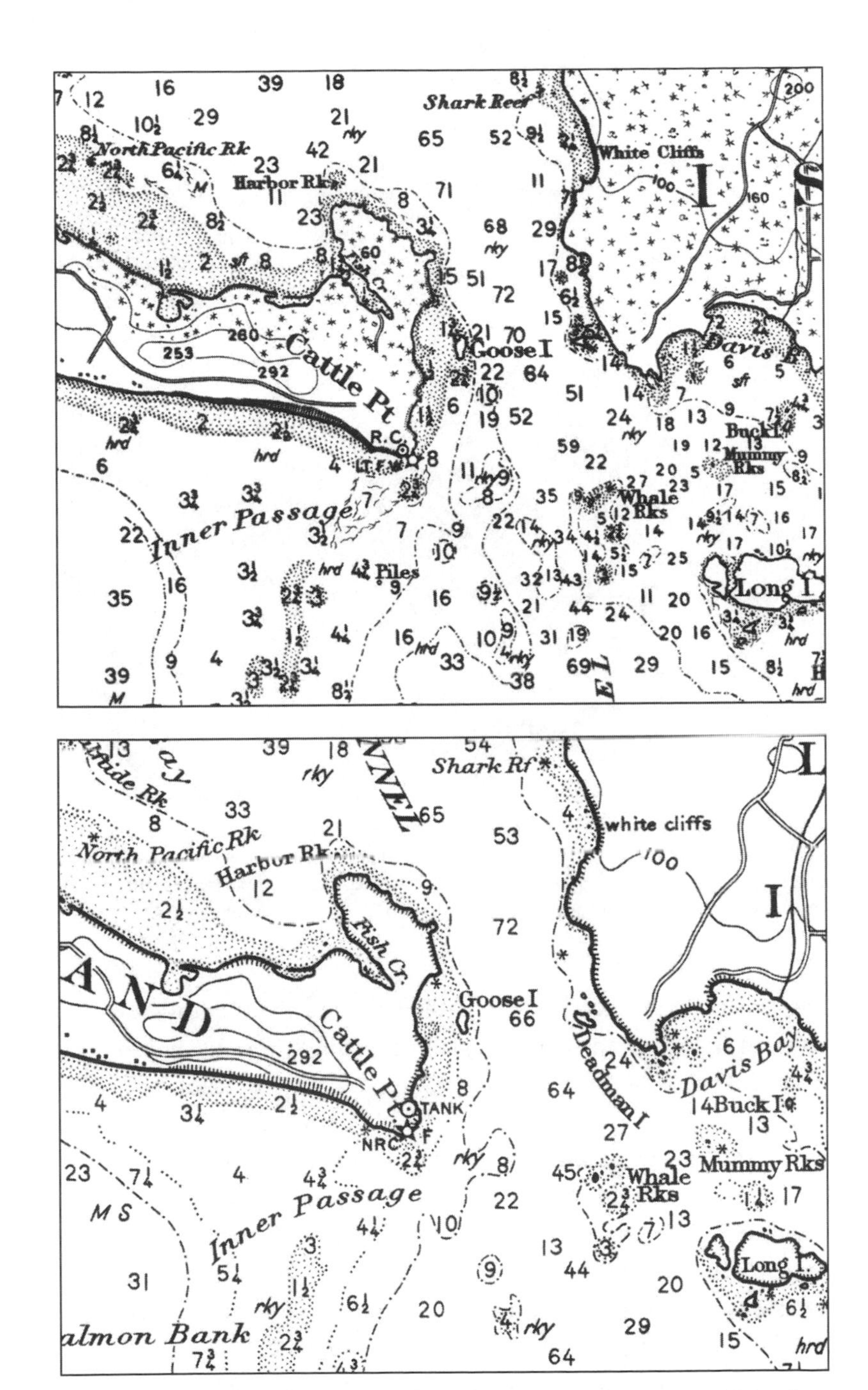

Figure 11.6. Excerpts from 1907 (*above*) and 1927 (*below*) editions of a 1:80,000 Coast and Geodetic Survey chart reflect adoption of a more open style, showing substantially fewer soundings and ignoring vegetation. From Deetz, *Cartography*, opp. 72.

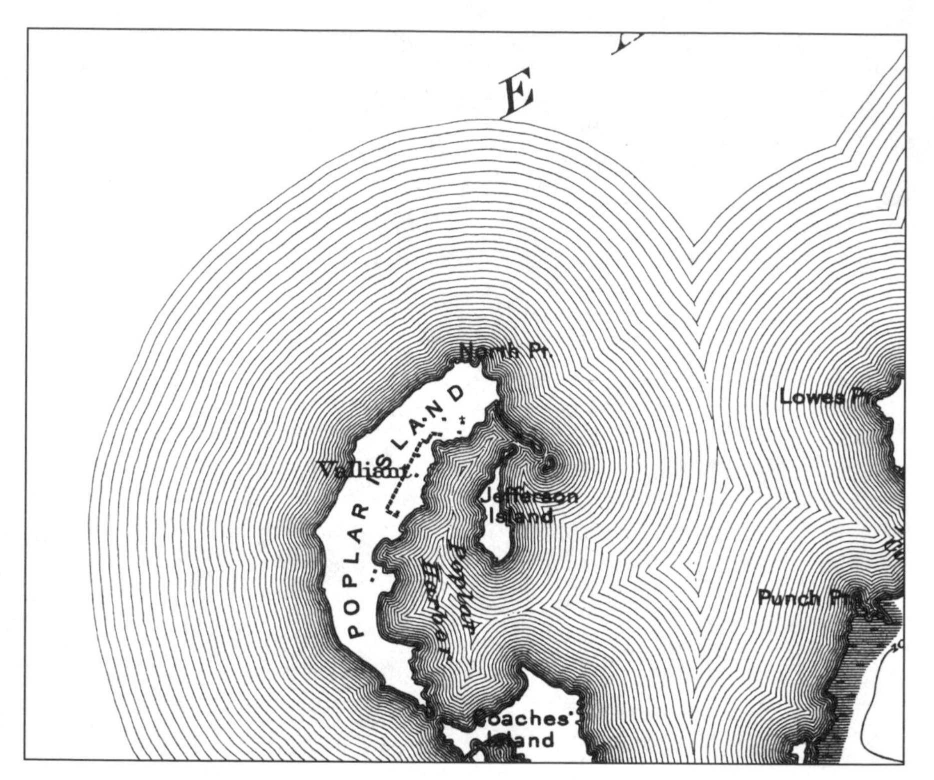

Figure 11.7. Before the U.S. Geological Survey adopted a solid blue tint for lakes, rivers, and the sea, its topographic maps highlighted the shoreline with a succession of thin blue lines progressively farther apart. From the U.S. Geological Survey's Annapolis, Md., 15-minute topographic map, published in 1904 at 1:62,500 and reprinted in 1944.

line directly south, less than a tenth of an inch away. Paralleling the coast on the seaward side, a single line of small, evenly spaced dots represents the low-water shoreline. Immediately adjacent, a zone of stipple or "sanding," studded sporadically with soundings like "2½," marks the extent of shallow water. While experienced chart users could quickly decode these seemingly arcane coastline symbols, novices no doubt appreciated the accelerated effort, starting in the early 1940s, to replace black-and-white sanding with a blue tint.[14]

Users of Geological Survey topographic maps had always enjoyed color-coding, but until the 1940s almost all USGS quadrangle maps sported only three colors: brown for relief, blue for water, and black for highways, boundaries, and other "culture."[15] Ponds, lakes,

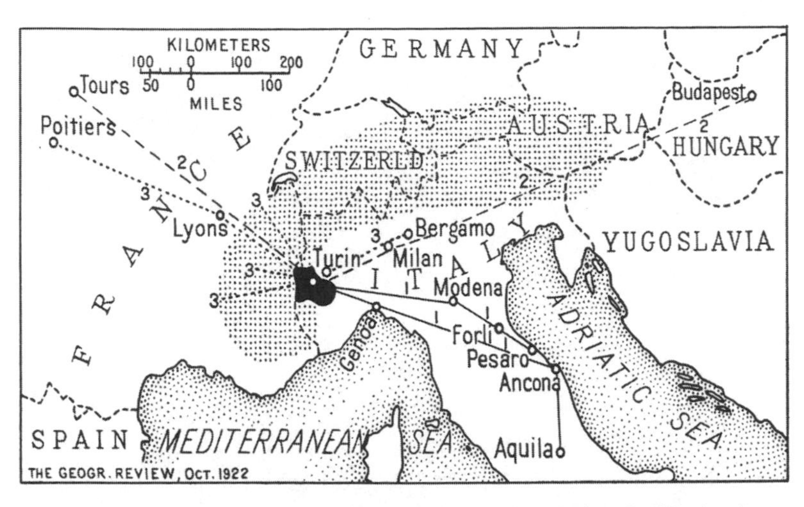

Figure 11.8. Map showing the "sphere of influence" of the fair at Pinerolo, Italy, for pigs (1), horses (2), and mules (3). Reduced to 86 percent of the original map's size in Allix, "The Geography of Fairs: Illustrated by Old-World Examples," 555.

reservoirs, bays, and larger water bodies were rendered with thin blue lines running parallel to the shoreline and spaced progressively farther apart. Mimicking a solid blue near the shore, the water lining (as it was called) petered out about an inch away from land. On the example in figure 11.7, from a 1904 map of Chesapeake Bay near Annapolis, I counted thirty-eight lines, except where water lining encountered the similar treatment for an opposing feature.[16] I chose this example to show the delicate details, which appear more severe in black-and-white.

Water lining was surprisingly long-lived. My copy of the Annapolis map, printed in 1944, perpetuates hydrographic symbols engraved four decades earlier. Although newer pre–World War II mapping reflected the present scheme—a light blue tint running from shore to shore plus a few soundings and depth contours picked up from nautical charts (see, for example, fig. 1.1)—USGS frugally recycled these fascinating copperplate engravings until new surveys, often at a larger, more detailed scale, demanded new artwork.

Throughout much of the twentieth century the more stylish small-scale thematic maps in newspapers, textbooks, and academic journals portrayed water bodies with a coastal stippling lifted from

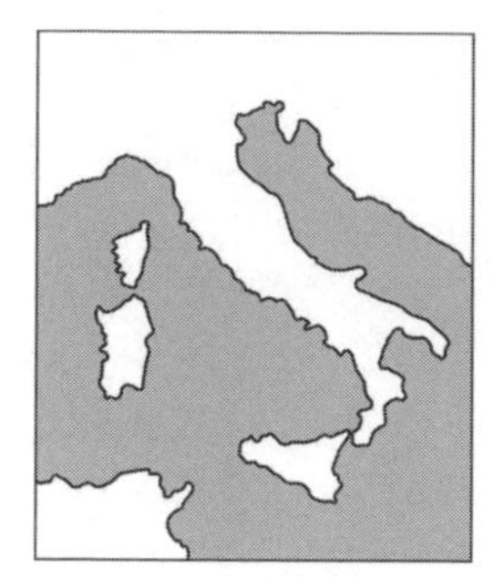

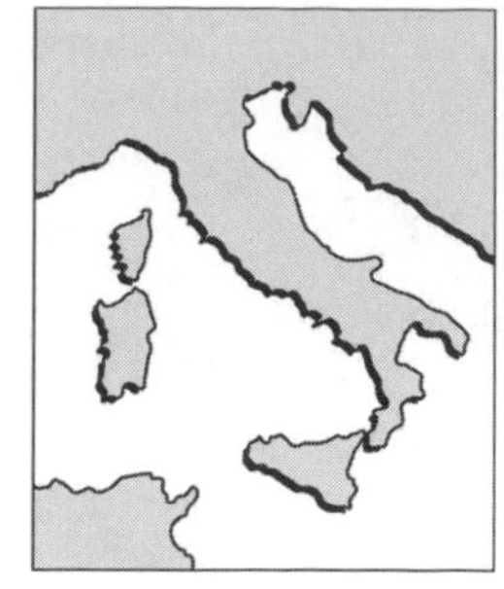

Figure 11.9. Three examples of how graphic layering with illustration software can readily differentiate land and water.

black-and-white nautical charts. A good example is the *Geographical Review*'s treatment of Mediterranean and Adriatic waters in a 1922 map drawn to illustrate the extraordinary attraction of a medieval market fair in Pinerolo, in the Italian Alps (fig. 11.8).[17] Dots were placed on the drawing individually, by hand, with larger, more closely spaced dots creating a darker symbol near the shoreline. As I recall from cartography lectures at Penn State, where I went to graduate school, the goal was an "irregular irregularity" in which dots faintly mimicked the trend of the coast and blended seamlessly with hand-lettered feature names.[18] My instructor was George Deasy, who honed his mapmaking skills at the War Department during World War II, and over subsequent decades compiled a list of sixty-seven rules, one warning against overly uniform stippling and another advising careful placement of hemispherical drops of ink, which dried to form perfectly circular dots. Fortunately for his students, Deasy allowed photomechanical type and wax-backed sheets of Zip-A-Tone, from which graytone water symbols could be cut with an X-acto knife—technical shortcuts that changed the look of manual cartography.

Nowadays geographers and newspaper artists easily create graytones and other "fill patterns" with illustration software like Adobe Illustrator, which treats the map as a series of layers. Mapmakers use electronic layering to differentiate land and water in diverse ways, as demonstrated with the small-scale black-and-white map of Italy and surrounding waters in figure 11.9. The left panel, in which the Mediterranean and its appendages are a medium gray, was composed as a graphic sandwich with four layers: a gray rectangle on the bottom, a collection of opaque white "objects" to block out the gray where there's land, a solid black coastline, and a thinner frame called a "neat line" on top. The gray water emphasizes the distinctive shape of the Italian coastline, while the white land provides a high-contrast background for symbols representing boundaries, cities, and transport routes. In the center panel, which also has four layers, a grid of meridians and parallels replaces the gray rectangle as the lowermost layer. The grid lines appear to run beneath the land, thereby creating distinct visual planes for areas above and below the shoreline. The right-hand panel asks the viewer to imagine a three-dimensional map illuminated from the top right.[19] The land layer, which is light gray rather than opaque white, mostly covers an identical but solid-black image of the land that has been shifted slightly to the left and bottom to form a so-called drop shadow. In the 1980s, when the Apple Macintosh and laser printer spurred a dramatic increase in newspaper maps, exaggerated drop shadows became a graphic cliché, which the more aesthetically astute journalistic cartographers quickly abandoned.

These three maps work particularly well because of the Italian coastline's distinctive bootlike shape. Other well-known coastlines that trigger instant land-water discrimination, at least among North American map viewers, are Chesapeake Bay and the Florida Peninsula. At the larger, more detailed scale of the harbor chart, the short, straight lines of port facilities invite immediate recognition as man-made features, largely because (to recite the mantra of landscape gardeners) nature abhors straight lines. Complex configurations can be similarly revealing. For example, few people can examine a topographic map of Cape Coral, Florida, without seeing its maze of canals as an artificial shoreline carved out of a mangrove swamp and cleverly crafted to maximize the number of waterfront lots. Of course, the story is helped along by a vine-like pattern of streets with names like Dolphin Drive and De Soto Court.

One of my favorite cartographic narratives is the Manhattan shoreline as mapped by Egbert Ludovickus Vielé, a West Point graduate and Union Army general who was New York City's chief

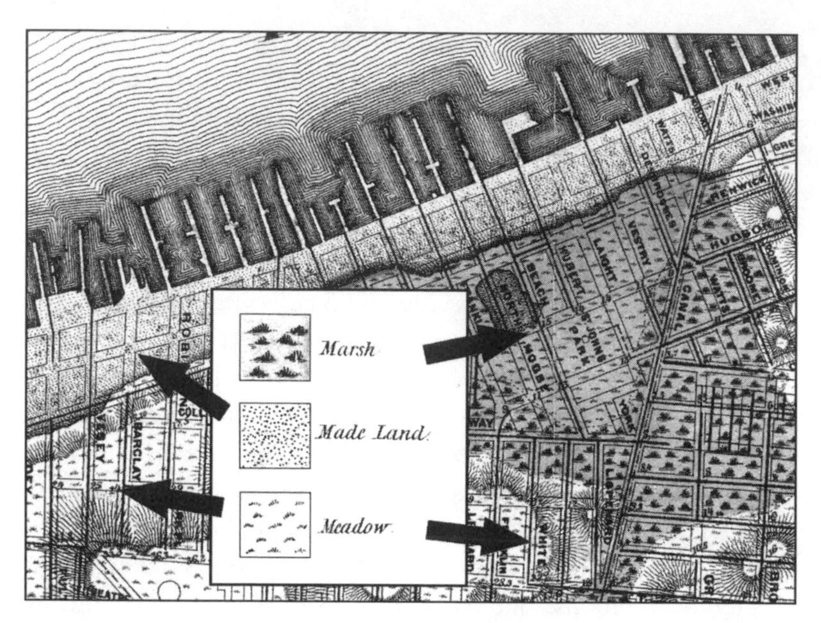

Figure 11.10. Tribeca as shown on Egbert Ludovickus Viele's Sanitary and Topographical Map of the City and Island of New York, printed and hand-colored in 1865. Excerpt from the map key shows the symbols for marsh (greenish blue), "made land" (light orange), and meadow (faint blue). From Citizens' Association of New York, Council of Hygiene and Public Health, *Report of the Council of Hygiene and Public Health of the Citizens' Association of New York upon the Sanitary Condition of the City.*

engineer. Viele was a member of the Council of Hygiene and Public Health of the Citizens' Association of New York, which in 1865 published a 360-page report that included a folded copy of his lithographed, hand-colored map.[20] Laid out in landscape format, the Topographic and Sanitary Map of the City and Island of New York placed north to the right and the Hudson River at the top. As revealed in an excerpt covering what is now Tribeca, on the Lower West Side (fig. 11.10), much of the city's waterfront is "made land," a quaint term for the dirt, garbage, and other debris dumped and compacted inside the bulkhead line—a sort of legal shoreline—running along West Street, perpendicular to the Hudson River wharfs. The connotation of *sanitary* in the map's title is further apparent in the thick black lines down the center of most streets. These heavy lines represent sanitary sewers, buried beneath the street and configured

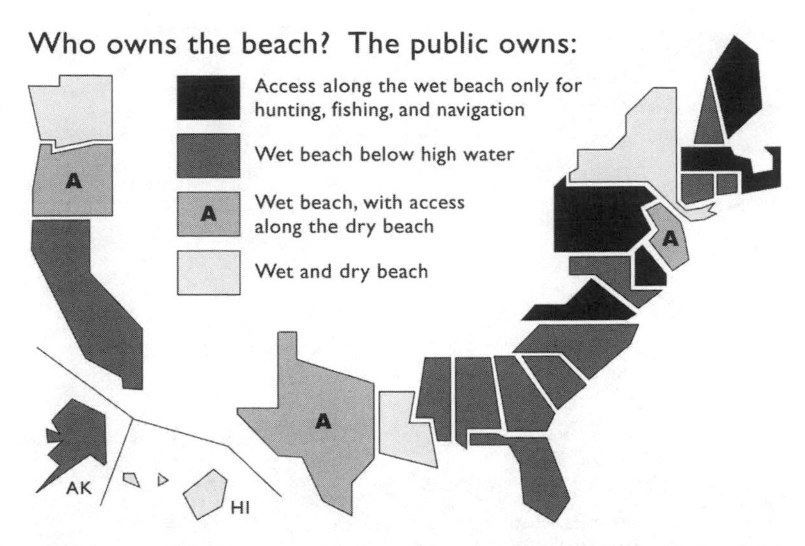

Figure 11.11. Public rights to beach access, by state. In general, the most restrictive states are Maine, Massachusetts, Pennsylvania, Delaware, and Virginia, which recognize a public right below the low-water line in addition to limited access to tidal lands (between high and low water) for hunting, fishing, and navigation. By contrast, Washington, Hawaii, Louisiana, and New York recognize public ownership of both the wet beach (between mean high water and mean low water) and the dry beach (between mean high water and the vegetation line, farther inland). In Oregon, Texas, and New Jersey the public owns the wet beach but also has the right to use the dry beach. In California, Alaska, New Hampshire, Rhode Island, Connecticut, Maryland, and six southeastern states, citizens also have a legal right only to the wet beach. Redrawn from a map in Titus, "Rising Seas, Coastal Erosion, and the Takings Clause: How to Save Wetlands and Beaches without Hurting Property Owners," 1367.

to transport fecal matter into the harbor without what contemporary city officials politely call "wastewater treatment."

Despite the coastline's historic convenience as a landfill and cesspool, beach walking can be a refreshing experience for anyone priced out of the market for waterfront housing. But getting onto the beach is not as straightforward as it once was, thanks to an epidemic of seaside retirement enclaves and trophy homes. Coastal homeowners don't want outsiders traipsing through their side yards, and a few even resist citizens' legal use of the narrow strip of public land between private holdings and the seabed.[21] Because maps are often the key to beach access, my final exhibits are a national map describing differences among the states in common-law

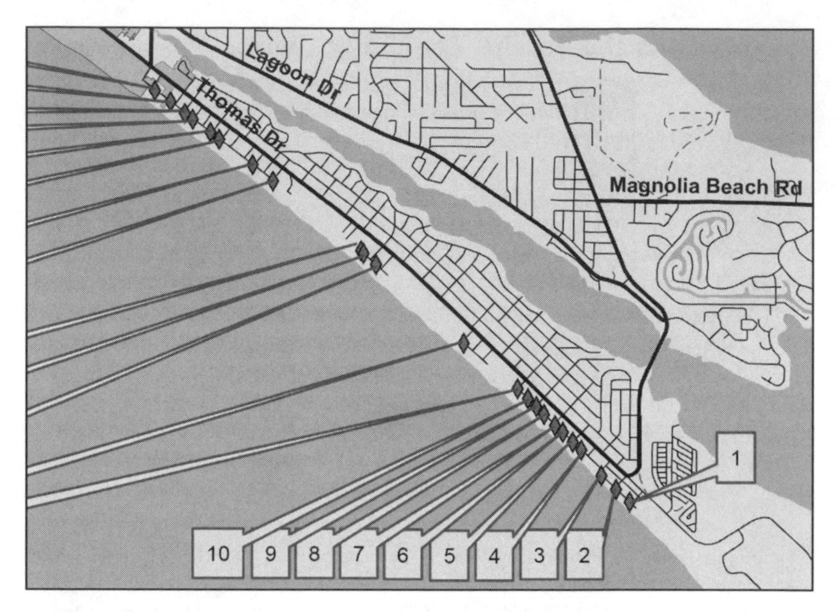

Figure 11.12. Portion of Bay County, Florida's online beach access map, which can be downloaded from the county Web site as a zoomable .pdf file. Numbers represent access points, described in an accompanying table that lists their street address, access width, and type of boardwalk, if any. From Bay County, Florida, Leisure Services Division, Beach Access Points Map.

beach access (fig. 11.11) and an exemplary local map pinpointing numerous beach-access points in Bay County, Florida (fig. 11.12).[22] The cartographic coastline, which started out as a navigation tool for mariners and fishermen, now helps beachcombers, surfers, birders, sunbathers, and families with small children enjoy their shoreline heritage.

TWELVE

EPILOGUE

Different kinds of cartographers map the coastline for different reasons, and each community of mapmakers has developed its own standards to ensure a consistent, reliable product. These standards have changed over time, largely because of changes in user requirements and mapping technology, and new uses and new technologies have fostered the emergence of four distinct types of cartographic coastline: a high-water line visible from offshore, a low-water shoreline to anchor bathymetric contours important to safe navigation, a storm-surge line that supports evacuation planning and flood insurance, and an inundation line describing the plausible effect of climate change and rising seas.

The first two cartographic coastlines are closely related to marine navigation. The oldest is the high-water line, an indispensable feature on any sailing chart and vital to national defense. In the early nineteenth century, trigonometric and systematic topographic surveys provided comparatively detailed, large-scale coastal charts useful for coastwise shipping. This first shoreline was complemented later in the century by a low-water line that reflected the hazards of coastal waters to larger cargo vessels as well as advances in tide prediction and hydrographic surveying. The low-water coastline required the conscious selection of a standardized definition, typically chosen from among mean low water, mean lower-low water, or lowest astronomical tide. NOAA's predecessors initially used mean low water for the Atlantic and Gulf coasts, but adopted mean lower-low water for the Pacific coast, where the successive low tides can be radically different. In 1980 the latter definition became the national standard for all three coasts, but only for civilian charts. By contrast, the American military, like the British Admiralty, references its soundings to the lowest astronomical tide.

Comparatively recent, the last two cartographic coastlines reflect the opportune coincidence of computer modeling and a growing awareness during the latter half of the twentieth century that densely populated shorelines pose a risk to life and property. In addition to a wider range of available definitions, these third and fourth shorelines embody a set of scientific assumptions or societal beliefs. Flood mapping, for example, demonstrates a collective faith in both the value of flood insurance and the need to regulate coastal development—how else to explain public acceptance of a statistical fantasy defined by a hypothetical flood with a recurrence interval of a hundred years? The third coastline can also be customized for specific events, as when the National Weather Service and state emergency managers base evacuation orders on storm-surge models tailored to the predicted path and severity of an approaching hurricane. Cartographic forecasts of sea level rise are even more flexible, given the need to choose a time horizon as well as make assumptions about climatic shifts, anthropogenic warming, and the behavior of ice sheets.

Coastal change is a persistent, increasingly visible environmental problem as land development destroys coastal wetlands and other natural buffers, which in turn magnifies the impact of coastal storms. Despite uncertainty arising from incomplete data, unavoidably simplified computer models, and the limited predictability of complex nonlinear systems like the atmosphere and the oceans, inundation maps are valuable tools for public officials who must plan for and respond to coastal catastrophes, including long-term shifts in the land-sea boundary that can inflict trillions of dollars of property damage and displace millions of people.[1] Hypothetical coastlines can also help private citizens make better short-run personal decisions about where to live and when to evacuate as well as better understand natural forces with grave consequences for future generations. Although skeptics see wide differences among the simulated inundations of climate change scenarios as an excuse for doing nothing, the need for substantial shifts in public policy is apparent in powerful cartographic comparisons like the coastlines in figures 10.9 and 10.10, which document the progressive retreat of the unarmored barrier island south of Ocean City, Maryland, over eighty years of rising seas.

The only certainty in plotting Type IV cartographic coastlines is the need for reliable elevation data; all else is conjecture—but necessary conjecture because of the dire consequences of ignoring a readily apparent and steadily escalating catastrophe. The key questions, of course, concern the plausible height and rate of sea level rise and the timeliness and effectiveness of human response. How well Earth's inhabitants will respond is very much a question of how imminently they see the threat. A map showing much of the planet's lowlands under water is readily undermined if its year date of 2500 becomes an excuse for deferring action on coastal settlement and greenhouse gasses, especially if another scientifically credible map suggests the inundation will occur in, say, 2400 or 2600. I'm not optimistic that humankind will adopt an ethic of stewardship that reaches more than a couple generations into the future. As our species and its political institutions have evolved, big issues invite procrastination.

Provocative inundation maps can help us recognize the seriousness of climate change. For me the most riveting images in Al Gore's book *An Inconvenient Truth: The Planetary Emergency of Global Warming and What We Can Do About It*, a spin-off from his Oscar-winning movie of the same name, is a twelve-page series of comparative, now-and-later inundation maps illustrating the effect of a 20-foot (6.1 meter) rise on Florida, San Francisco Bay, the Netherlands, the Chinese megacities of Beijing and Shanghai, Bangladesh and Calcutta, and lower Manhattan.[2] A clever propagandist—I intend this as a compliment—the former vice president sidestepped the scientist's compulsion to tag the maps with a year date and reinforced the visual shock by quoting British science advisor David King's observation that "the maps of the world will have to be redrawn."[3] Although impressed by Gore's graphics, I'm disappointed that he did not include equally dramatic maps of coastal flooding and shoreline erosion—urgent short-term problems that mirror the more violent impacts of sea level rise. The public needs to recognize the Type III cartographic coastline as the precursor of its Type IV counterpart.

Because effective response to sea level rise demands local recognition of the potential effects on people's neighborhoods, nearby roads and parks, and local assets such as workplaces, stores, and historic sites, I am intrigued by the Light Blue Line initiative (www.lightblueline.org), launched in 2006 by community activists in Santa Barbara, California. As the name vaguely implies, their goal is to paint a seven-meter elevation contour across streets, parking lots, roofs, and interior floors; roughly equivalent to Al Gore's twenty-foot shoreline, seven meters represents a complete meltdown of the Greenland ice sheet, or a plausible worst-case reduction of both the Greenland and West Antarctic ice sheets. Although transferring this hypothetical Type IV shoreline from map to landscape is a jarringly dramatic gesture, its success requires adoption by other coastal communities as well as a stubborn commitment not only to repaint the line every year or so but also to adjust the line's position occasionally to reflect better elevation data and a fuller understanding of climate change. It's simpler and perhaps no less effective to store the blue line electronically on Google Earth, where it's assured a wider audience.

NOTES

CHAPTER 1

1. For a description of the imagery, see Tucker, Grant, and Dykstra, "NASA's Global Orthorectified Landsat Data Set."

2. Shalowitz, *Interpretation and Use of Coast and Geodetic Survey Data*, 479–85. This book is volume 2 of a three-volume set titled *Shore and Sea Boundaries*. Shalowitz, who produced the first two volumes, died in 1975. Volume 3, published in 2000 in a different format, summarizes more recent judicial rulings and international agreements.

3. Shalowitz provided no further description, but Curt Loy, the current NOAA boundary expert, confirmed this interpretation. Curtis Loy (Office of Coast Survey, NOAA headquarters), conversation with author, April 11, 2005. The device is sometimes called a map measurer, a curvimeter, or an opisometer. See Maling, "How Long Is a Piece of String?" A 1975 flyer faxed from the Office of Coast Survey in 2005 indicates that NOAA still accepts the 88,633-mile figure for the United State proper, exclusive of the Great Lakes. The flyer did not mention 95,000 miles or a nonrounded equivalent. See National Oceanic and Atmospheric Administration, *The Coastline of the Unites States*.

4. Cynthia Fowler (GIS Integration and Development Program, Coastal Information Services Branch, NOAA Coastal Services Center), e-mail communication with author, April 5, 2005; and Curtis Loy (Office of Coast Survey, NOAA Headquarters), conversation with author, April 11, 2005.

5. Shalowitz, *Interpretation and Use*, 483.

6. Ibid., 483–84.

7. Linklater, *Measuring America*, 71–73.

8. Shalowitz, *Interpretation and Use*, 479–83. Re-measurement in 1948 largely reconfirmed the original measurements, published in 1915. See U.S. Coast and Geodetic Survey, *Lengths, in Statute Miles, of the General Coast Line and Tidal Shore Line of the United States and Outlying Territories*. California's coastline shrank from 913 to 840 miles, perhaps because of improved charting, while

Mississippi's increased from 14 to 44 miles, to correct an apparent blunder. In 1961 a fresh look at the Florida Keys increased that state's general coastline from 1,197 to 1,350 miles.

9. U.S. Department of State, *Sovereignty of the Sea*, 16–20.

CHAPTER 2

1. Tide predictions for figure 2.1 are from National Oceanic and Atmospheric Administration, "Predicted Water Level Data."

2. Hiscock, *Cruising under Sail*, 309.

3. Conway, "Origins of the Terms 'Spring' and 'Neap' Tides."

4. For a concise introduction to the astronomical and oceanographic influences on tides, see Clancy, *The Tides: Pulse of the Earth*.

5. Zahl, "The Giant Tides of Fundy."

6. According to Mike Bonin of the Communications Branch in the Maritimes Region of the Canadian Coast Guard, the cradle shown in figure 2.3, also referred to as a crib, is a necessary adaptation to the bay's large tides. Mike Bonin, e-mail communication, February 8, 2007.

7. Shalowitz, *Interpretation and Use of Coast and Geodetic Survey Data*, 254–55.

8. Calder, *How to Read a Nautical Chart*, 50; and Mills and Gill, "Water Levels and Flow," esp. 267–68.

9. A tidal datum similar to mean lower-low water was already prescribed for much of the Gulf coast. Hicks, *The National Tidal Datum Convention of 1980*, 17–18.

10. Tide statistics for this paragraph are from National Oceanic and Atmospheric Administration, "Predicted Water Level Data."

11. Wheeler, *A Practical Manual of Tides and Waves*, 162–63.

12. Quoted in ibid., 161.

13. Cartwright, *Tides*, 97–100.

14. Quoted in Wheeler, *Practical Manual*, 161.

15. U.S. Coast and Geodetic Survey, *Description of the U.S. Coast and Geodetic Survey Tide-Predicting Machine No. 2*, 12–13, 34.

16. Ibid., 13–35; National Oceanic and Atmospheric Administration, Center for Operational Oceanic Products and Services. "Tide-Predicting Machine No. 2"; and Schureman, *Manual of Harmonic Analysis and Prediction of Tides*, 126–47.

17. Jones, *Elements of Chart Making*, 13.

18. Ibid.

19. Todd Ehret (Physical Oceanographer, NOAA Center for Operational Oceanographic Products and Services), conversation with author, April 11, 2005.

20. Cartwright, *Tides*, 106; and Mays, "Thorough and Versatile Tide Prediction by Computer." A phase-in of digital modeling began in 1956. Harris, Pore, and Cummings, "Tide and Tidal Current Prediction by High Speed Digital Computer." An improved replacement for the analog machine was "75% completed" as of 1959, but apparently never used. See Parkhurst, "Salient Features of the Design of a New Tide-Predicting Machine," quote on 107.

21. Hess, "Tide Datums and Tide Coordination."

22. Calder, *How to Read a Nautical Chart*, 68, 90, 132–33. Even though the instantaneous, wave-blurred boundary on an air photo is not an exact delineation, tide-coordinated imagery can point out its approximate position. See Graham, Sault, and Bailey, "National Ocean Service Shoreline—Past, Present, Future," 19, 24.

23. According to Morris Thompson, who authored an official description of USGS products, "The mean high water line is shown as the shoreline on nautical charts of the National Ocean Survey [sic] and topographic maps of the Geological Survey." Thompson, *Maps for America*, 43. A worthy goal perhaps but not easily met. According to a recent National Research Council study, "the shoreline delineated on USGS topographic maps is derived from stereo aerial photographs without reference to tidal (MHW or MLLW) datums." National Research Council, *A Geospatial Framework for the Coastal Zone*, 45.

24. National Geodetic Survey, "Importance of Shoreline."

25. Because of historic boundaries predating statehood, Texas and Florida have jurisdiction over a relatively wide coastal zone, which extends three leagues, rather than three nautical miles, into the Gulf of Mexico. See De Vorsey, "Florida's Seaward Boundary." A nautical mile is the length of one minute of longitude at the equator (1.151 statute miles or 1.852 km), and a league is three nautical miles.

26. Subtle distinctions preclude complete agreement on the classification of the legal shoreline defined for Hawaii and Louisiana. See Minerals Management Service, *Implementation Plan for a Multipurpose Cadastre*; and National Geodetic Survey, "Importance of Shoreline" (diagram).

27. U.S. Army Corps of Engineers, Interagency Performance Evaluation Task Force, *Performance Evaluation of the New Orleans and Southeast Louisiana Hurricane Protection System, Draft Final Report, 1 June* 2006, vol. 1—Executive Summary and Overview, esp. ch. 1, p. 5.

28. Calder, *How to Read a Nautical Chart*, 50–51.

29. Ibid., 120.

30. The "metonic cycle" is actually 18.6 years long, but upward rounding to nineteen years obviates a seemingly arbitrary start date and is sufficiently long to dampen meteorological influences. National Oceanic and Atmospheric Administration, *Computational Techniques for Tidal Datums Handbook*, 16–20. A new tidal epoch is initiated roughly every twenty years.

31. Trends in mean sea level are from National Oceanic and Atmospheric Administration, "Sea Levels Online."

32. Weber, *The Coast and Geodetic Survey*, 1–5.

33. Graham, Sault, and Bailey, "National Ocean Service Shoreline," quote on 18. Also see Pajak and Leatherman, "The High Water Line as Shoreline Indicator."

34. Swainson, *Topographic Manual*, 69.

35. Graham, Sault, and Bailey, "National Ocean Service Shoreline," 26–27.

36. Leatherman, "Shoreline Mapping," 31.

37. U.S. Coast Survey, *Coast-Pilot, 1874*, 495 (emphasis mine).

38. Lynne Jones (president, Georgetown Historical Society), letter to author, March 31, 2005.

39. Federal Emergency Management Agency, "Multi-Year Flood Hazard Identification Plan"; and Lowe, "Federal Emergency Management Agency's Multi-Hazard Flood Map Modernization and *The National Map*."

CHAPTER 3

1. For an overview of the map and its acquisition, see Hebért, "The Map That Named America"; and Hessler, "Warping Waldseemüller: A Phenomenological and Computational Study of the 1507 World Map."

2. Anderson, "A Quest to Claim America's Birth Certificate."

3. Most writers assign the rediscovery to 1901, but Joseph Fischer, the Jesuit scholar who found the map at Wolfegg Castle, reported the date as 1900. See Fischer and Wieser, introduction to *The Cosmographiae Introductio of Martin Waldseemüller in Facsimile,* 16.

4. Suárez, *Early Mapping of the Pacific*, 37

5. Ibid., 38.

6. For descriptions, see Harris, "The Waldseemüller World Map"; and Heawood, "The Waldseemüller Facsimiles," 767–69. Harris argues that the surviving copy of the 1507 world map was printed in 1515 or later.

7. Frey, "Donor's Gift Is New-World Class."

8. Polk, *The Island of California,* 105–20.

9. Briggs, "A Treatise of the Northwest Passage," 48; quoted in Polk, *The Island of California,* 284.

10. Polk, *The Island of California*, 283–86; and Tooley, *California as an Island,* 3.

11. Polk, *The Island of California,* 269–76; quotation on 275.

12. Ascensión, "A Brief Report of the Discovery in the South Sea," 110; quoted in Polk, *The Island of California,* 275.

13. Polk, *The Island of California,* 286–94.

14. Ibid., plate 44 on 287.

15. Tooley, *California as an Island.* Understandably this cartographic curiosity attracted considerable attention from scholars in California; for examples, see Leighly, *California as an Island,* and McLaughlin and Mayo, *The Mapping of California as an Island: An Illustrated Checklist.*

16. Campbell, "Portolan Charts"; and Loomer, "Mathematical Analysis of Medieval Sea Charts."

17. Tooley, *California as an Island*, 4.

18. Polk, *The Island of California*, 297–302.

19. Tooley, *California as an Island*, 4.

20. Ibid., 23.

21. Polk, *The Island of California*, 324–25.

22. I could not verify the decree, which is widely quoted, for example, in Polk, *The Island of California*, 326; and Tooley, *California as an Island*, 4.

23. Morison, "Texts and Translations of the Journal of Columbus's First Voyage," 237; and Thatcher, *Christopher Columbus: His Life, His Works, His Remains*, 1:477.

24. Facsimile of the map is a foldout, between pages 4 and 5, in Alba, *Nuevos Autógrafos de Cristóbal Colón y Relaciones de Ultramar.*

25. See, for example, Burden, *The Mapping of North America: A List of Printed Maps, 1511–1670*, xv; and Nebenzahl, *Atlas of Columbus*, 26.

26. Thatcher, *Christopher Columbus*, map on 3:89; quotation on 3:94.

27. Morison, "Route of Columbus along the North Coast of Haiti and the Site of Navidad," 260, 263.

28. See Bargar, B. D. "Samuel Eliot Morison"; and *Encyclopædia Britannica Online*, s.v. "Morison, Samuel Eliot," http://www.search.eb.com/eb/article?tocId=9053759 (accessed April 25, 2005).

29. Morison, "Route of Columbus," quotations on 239, 260, and map between 267 and 268.

30. Morison, "Route of Columbus," 240.

31. Ibid., 250.

32. Ibid., quotations on 240, 264. Morison's latter criticism refers specifically to the coast between the present-day settlements of Petite Anse and Fort Liberté.

33. Ibid., 262.

34. Ibid. The juxtaposition includes an enigmatic "Map of 1516," identified elsewhere in the article as the "Bologna map." Ibid., 267. Juan de la Cosa, the cartographer on Columbus's second voyage, is sometimes confused with Juan de la Cosa, the master of the ill-fated *Santa Maria*. See Morison, *Admiral of the Ocean Sea*, 1:186–87. The former Cosa later crossed the Atlantic with Vespuci. For reproductions and concise analyses of his map, see Bagrow, *History of Cartography*, 107–9, plate LVI; and Nebenzahl, *Atlas of Columbus and the Great Discoveries*, 30–33.

35. For discussion of the perils of map compilation from coastal sketches, see Skelton, "The Cartography of Columbus' First Voyage," esp. 223.

36. For discussion of positioning methods used by (or at least available to) Columbus, see Collinder, *A History of Marine Navigation*, 123–29; and Morison, *Admiral of the Ocean Sea*, 1:242–55.

37. *Encyclopedia of Astronomy and Astrophysics*, ed. Paul Murdin (London: Institute of Physics, 2001), s.v. "Gemma Frisius, Reiner."

38. Macey, *The Dynamics of Progress*, 11.

39. Taylor and Richey, *The Geometrical Seaman*, 87–89.

40. Schilder, "Willem Jansz. Blaeu's Wall Map of the World, on Mercator's Projection, 1606–07 and Its Influence"; Shirley, *The Mapping of the World*, 272–76; and Whitfield, *The Charting of the Oceans*, 69–74.

41. Robinson, *Marine Cartography in Britain*, 48; and Whitfield, *The Charting of the Oceans*, 71–74.

42. Schilder, "Willem Jansz. Blaeu's Wall Map of the World, on Mercator's Projection, 1606–07 and Its Influence," 39.

43. Ibid., 41. According the Schilder, the folio-size map in figure 3.6 is identical in design to a substantially larger wall map, printed in four sheets. Ibid., 50.

44. Sobel, *Longitude*.

45. The misplaced islands were found on the 1864 edition of Admiralty chart 2683. Stommel, *Lost Islands*, 8.

46. Ibid., 98.

47. "*The Times*" *Atlas*, 113.

48. Stommel, *Lost Islands*, 98–99; quotation on 98.

CHAPTER 4

1. U.S. Congress, House Committee on Naval Affairs, *Steam Communication with China and the Sandwich Islands*, 30th Cong., 1st sess., 1848, rept. 596.

2. Maury published his letter of transmittal to the report's author, Congressman T. Butler King, who chaired the House Committee on Naval Affairs. See Maury, "Steam Navigation to China."

3. Maury, "Blank Charts on Board Public Cruisers," 458.

4. Hearn, *Tracks in the Sea*, 106, 112–21; and Williams, *Matthew Fontaine Maury*, 178–95.

5. Weber, *Hydrographic Office*, 18.

6. For examples, see Guthorn, *United States Coastal Charts, 1783–1861*.

7. The 1807 legislation enabled NOAA (the National Oceanic and Atmospheric Administration) to celebrate a 200-year anniversary in 2007. For a concise history of the Survey of the Coast, its successor agencies, and their activities, see Cloud, "The 200th Anniversary of the Survey of the Coast."

8. Cajori, *The Chequered Career of Ferdinand Rudolph Hassler*, 37–41.

9. Bencker, "The Development of Maritime Hydrography and Methods of Navigation," 127; and Whitfield, *The Charting of the Oceans*, 118.

10. Theberge, *The Coast Survey: 1807–1867*, 5–6.

11. Konvitz, *Cartography in France, 1660–1848*, esp. 7–31.

12. My greatly simplified description of geodetic surveying ignores steps for determining the direction of true north at control stations and for measuring the variation of magnetic north from true north. For a concise introduction to geodesy, see Smith, *Introduction to Geodesy: The History and Concepts of Modern Geodesy*; and U.S. Defense Mapping Agency, *Geodesy for the Layman*, esp. 9–12.

13. Hassler, "Papers on Various Subjects Connected with the Survey of the Coast of the United States."

14. Theberge, *The Coast Survey: 1807–1867*, 20.

15. Ibid., 5–27.

16. Hassler, "Papers," 246.

17. Ibid., 296.

18. Theberge, *The Coast Survey: 1807–1867*, 10.

19. Hassler, "Papers," 371–85.

20. Troughton's prowess as an instrument maker rested on his development of a mechanical dividing engine, used to partition circular scales into small, precise gradations. See Cox, "The Development of Survey Instrumentation, 1780–1980," 235–36.

21. Hassler, "Papers," 384.

22. Ibid., 385–408.

23. Ibid., 385–86.

24. Ibid., 273.

25. For an inventory of the apparatus, see ibid., 247.

26. Ibid.

27. Nelson and Dracup, "A Tale of Two Eras in American Geodesy," 166–67.

28. Cox, "The Development of Survey Instrumentation, 1780–1980," 245. Also see Simmons, "Geometric Techniques in Geodesy," 5.

29. Ibid., 348.

30. Wainwright, *Plane Table Manual*, 2.

31. The plane table designated for use by the Coast Survey was heavier, sturdier, and more cumbersome than those for other uses. Wilson, *Topographic, Trigonometric and Geodetic Surveying*, 146–64.

32. Jones, *Elements of Chart Making*, 8–9.

33. Ibid., 10.

34. In some situations, two party members would read separate angles between signals, and a recorder entered these angles and the depth reading in a notebook for later plotting ashore. See ibid., 8.

35. Weber, *The Coast and Geodetic Survey, 16–23*. Early in the twentieth century, hydrographers wary of pinnacle rocks and other dangerous upward projections from the seabed supplemented their linear sequences of soundings (as in fig. 4.8) with a "wire drag," a systematic method of trolling for shallow features with a wire supported at a fixed depth between two boats. Jones, "The Evolution of the Nautical Chart," 226–27.

36. Because they sampled sediments while relying on triangulated positions based on comparatively precise topographic and geodetic measurements, the hydrographic surveyor was, in as many ways as possible, the bottom feeder of coastal charting.

37. Cajori, *The Chequered Career of Ferdinand Rudolph Hassler*, 164–65, 177.

38. Ibid., 165–68; quotation on 168, from F. R. Hassler, *Principal Documents Relating to the Survey of the Coast of the United States* (New York: W. Van Norden, 1834), 101.

39. Cajori, *The Chequered Career of Ferdinand Rudolph Hassler*, 175–81; and Theberge, *The Coast Survey: 1807–1867*, 43–45.

40. Weber, *The Coast and Geodetic Survey*, 3–4.

41. The first of these interim maps was an 1835 chart of Bridgeport Harbor, Connecticut. Shalowitz, *Interpretation and Use of Coast and Geodetic Survey Data*, 13. The agency acquired a color press around 1900. Until then color, when used at all, was applied by hand. National Oceanic and Atmospheric Administration, Office of Coast Survey, "Historical Maps and Charts – Background."

42. Theberge, *The Coast Survey: 1807–1867*, 83–85.

43. The chart was printed in six sheets, priced at twenty-five cents each. In 1845 the agency offered a smaller, single-sheet version, published at 1:80,000, for seventy-five cents. Guthorn, *United States Coastal Charts, 1783–1861*, 71–72; and Shalowitz, *Interpretation and Use of Coast and Geodetic Survey Data*, 13.

44. Cajori, *The Chequered Career of Ferdinand Rudolph Hassler*, 233–34.

45. Hassler's manner no doubt alienated some colleagues and legislators. For examples of his idiosyncrasies, see Cajori, *The Chequered Career of Ferdinand Rudolph Hassler*, 46, 227; Glatthard and Bollinger, "Swiss Precision for U.S. Mapping: Ferdinand Rudolph Hassler, First Chief of U.S. Coast and Geodetic

Survey and U.S. Bureau of Standards," 5; Stachurski, "Longitude by Wire: The American Method," 16–17; and Theberge, *The Coast Survey: 1807–1867*, 67–71, 74–75.

46. Cajori, *The Chequered Career of Ferdinand Rudolph Hassler*, 236.

47. Nelson and Dracup, "A Tale of Two Eras in American Geodesy," 164.

48. Slotten, *Patronage, Practice, and the Culture of American Science*, 99.

49. Odgers, *Alexander Dallas Bache, Scientist and Educator, 1806–1867*, 176; Weddle, "The Blockade Board of 1861 and Union Naval Strategy"; Slotten, *Patronage, Practice, and the Culture of American Science*, 110.

50. Blockade Board, "Second Report of Conference for the Consideration of Measures for Effectually Blockading the South Atlantic Coast," 200.

51. For concise contemporary appraisals of these contributions, see Boutelle, "What Has the Coast Survey Done for Science?"; and Ogden, "The Survey of the Coast."

52. Schott, "Recent Contributions to Our Knowledge of the Earth's Shape and Size, by the United States Coast and Geodetic Survey."

53. Patton, "Recent Advancements in Coast and Geodetic Survey Methods," 5–6; and U.S. Coast and Geodetic Survey, *The United States Coast and Geodetic Survey: Its Work, Methods, and Organization*, 124–25.

CHAPTER 5

1. The notion of modes of mapping practice and cross-cutting enterprises originated with Matthew Edney, a coeditor of volume 4 of the *History of Cartography*, which addresses the eighteenth century. See Edney, "Cartography without 'Progress': Reinterpreting the Nature and Historical Development of Mapmaking."

2. The other mode that emerged in the twentieth century was dynamic cartography. Also significant was the emergence of academic cartography.

3. For an overview of early overhead imaging, by sketchpad and photograph, see Blachut and Burkhardt, *Historical Development of Photogrammetric Methods and Instruments*, esp. 4–10, 33–47; and Thompson and Gruner, "Foundations of Photogrammetry," esp. 3–11.

4. Wilson, *Topographic, Trigonometric and Geodetic Surveying*, 292–304.

5. Ibid., 295.

6. Collier, "The Impact on Topographic Mapping of Developments in Land and Air Survey: 1900–1939," 158–62.

7. Shalowitz, "Safeguarding Our Seaways," 257.

8. Shalowitz, *Interpretation and Use of Coast and Geodetic Survey Data*, 277.

9. Smith, *A History of Flying and Photography in the Photogrammetry Division of the National Ocean Survey, 1919–79*, 1, 10–18, 38.

10. The illustration from which figure 5.1 is an excerpt also appears in ibid., 39. A chronological set of project maps at the back notes coastal photography in 1927 along an eighty-mile strip of coastline near Naples. Ibid., 367. The revised map was probably compiled between January 1928 and July 1930.

11. U.S. Coast and Geodetic Survey, *Annual Report of the Director, United States Coast and Geodetic Survey to the Secretary of Commerce for the Fiscal Year Ended June 30, 1930*, 18–20.

12. Ibid., 18.

13. Reading, "Aerial Photography and Coast Surveys," 10.

14. Jones, "The Nine-Lens Air Camera in Use"; Reading, "The Nine Lens Aerial Camera of the Coast and Geodetic Survey"; and Tewinkel, "Stereoplotter for Nine-Lens Photographs."

15. For discussion of the sources of error that must be confronted when using aerial imagery, see Moore, "Shoreline Mapping Techniques."

16. Smith, *A History of Flying and Photography in the Photogrammetry Division of the National Ocean Survey, 1919–79*, 68.

17. According to a 1955 article by Robert W. Knox, assistant director of the Coast and Geodetic Survey, acquiring the air photos accounted for only 3 percent of the cost of making topographic maps from aerial imagery, while "the elaborate compilation process that the photograph must be subjected to in the office" accounted for the balance. See Knox, "Mapping the Earth," 73. An off-the-shelf aerial camera with an ultra-wide-angle lens able to capture a similarly broad scene replaced the nine-lens camera in 1961. Graham, Sault, and Bailey, "National Ocean Service Shoreline—Past, Present, and Future," 21.

18. The Coast and Geodetic Survey still used the plane table from time to time as late as the 1960s, primarily in areas with persistent cloud cover. Douglas Graham (Remote Sensing Division, National Geodetic Survey, NOAA headquarters), e-mail communication with author, July 7, 2005.

19. For further information on the generation and use of orthophotos, see Jensen, John R. "Issues Involving the Creation of Digital Elevation Models and Terrain Corrected Orthoimagery Using Soft-Copy Photogrammetry."

20. Leatherman, "Shoreline Mapping: A Comparison of Techniques."

21. For examples of the use of aerial imagery to map coastlines with diverse origins, see McCurdy, *Manual of Coastal Delineation from Aerial Photographs.*

22. Brooks, "Use of Aerial Photographs for Revision of Land Information on Nautical Charts." For a concise overview of the process of planning, making, and revising nautical charts, ca. 1955, see Jones, "Photogrammetric Surveys for Nautical Charts." As Jones notes, the Coast and Geodetic Survey developed and maintained a comparatively detailed topographic "base map," used in compiling published charts. Ibid., 111–12, 127. Occasionally the agency maintained two sets of photos: large-format, multilens imagery for topography and the high-water line, and a set of more detailed, lower-altitude single-lens imagery coordinated with the low-water line. Ibid., 114.

23. Jones, "Photogrammetric Surveys for Nautical Charts," 118–19.

24. Graham, Sault, and Bailey, "National Ocean Service Shoreline—Past, Present, and Future," 27–28. For a contemporary discussion of color photography's benefits, which include improved depth penetration, see Geary, "Coastal Hydrography."

25. Graham, Sault, and Bailey, "National Ocean Service Shoreline," 23–24. Also see Floyd, "National Ocean Service Shoreline Mapping Program."

26. Douglas Graham (Remote Sensing Division, National Geodetic Survey, NOAA headquarters), e-mail communication with author, July 11, 2005.

The four-meter guideline was established after August 13, 2002, when a high wave capsized a survey launch along the Alaskan coast killing crew member Eric Koss.

27. Richard P. Floyd (former chief of the NOAA Photogrammetry Division), telephone conversation with author, July 11, 2005.

28. For examples, see National Oceanic and Atmospheric Administration, National Geodetic Survey, Remote Sensing Division, "Imaging Spectroscopy (Hyperspectral)."

29. For examples, see Liu and Jazek, "A Complete High-Resolution Coastline of Antarctica Extracted from Orthorectified Radarsat SAR Imagery"; and National Oceanic and Atmospheric Administration, National Geodetic Survey, Remote Sensing Division, "Synthetic Aperture Radar (SAR)."

30. For examples, see Espey, "Using Commercial Satellite Imagery and GIS to Update NOAA ENCs"; National Oceanic and Atmospheric Administration, National Geodetic Survey, Remote Sensing Division, "CSCAP: Coast and Shoreline Change Analysis Program"; and Vidal, Graham, and Sault, "CSCAP: Coast and Shoreline Change Analysis Program—Using High-Resolution Satellite Imagery for Shoreline Change Evaluation within Ports." CSCAP also uses high-altitude aerial photography.

31. For a concise summary of lidar principles and products, see Flood, "Laser Altimetry: From Science to Commercial Lidar Mapping."

32. Example specification are from Florida International University, "Airborne Laser Technology."

33. Flood, "Product Definitions and Guidelines for Use in Specifying Lidar Deliverables."

34. For a comparison of bare-earth filtering techniques, see Zhang and Whitman, "Comparison of Three Algorithms for Filtering Airborne Lidar Data."

35. Wozencraft and Lillycrop, "SHOALS Airborne Coastal Mapping: Past, Present, and Future."

36. National Research Council, Committee on National Needs for Coastal Mapping and Charting, *A Geospatial Framework for the Coastal Zone: National Needs for Coastal Mapping and Charting*, 62–73, 114–15; quotation on 114.

37. Graham, Sault, and Bailey, "National Ocean Service Shoreline," 22; and Stockdon and others, "Estimation of Shoreline Position and Change Using Airborne Topographic Lidar Data," 510. Lidar can also be used with tide-coordinated orthophotography to automate delineation of the high-water line; see Robertson and others, "Mapping Shoreline Position Using Airborne Laser Altimetry."

CHAPTER 6

1. Mariners using electronic navigation systems ought not be surprised if the display screen places their gently rocking, bobbing craft several feet inland, above the shoreline. See Calder, *How to Read a Nautical Chart*, 19–21; and Perugini, "Behind the Accuracy of Electronic Charts."

2. For an illustrated description, see Chapman, *Piloting, Seamanship, and Small Boat Handling*, 380–81.

3. Monmonier, *Rhumb Lines and Map Wars*, 63–68, 79.

4. Snyder, *Flattening the Earth*, 48, 59–60. Franklin prepared his map in collaboration with his cousin, whaling captain Timothy Folger.

5. Ibid., 117–19.

6. Hassler's biography is rich with examples of his idiosyncrasies; see Cajori, *The Chequered Career of Ferdinand Rudolph Hassler.*

7. Hassler, "Papers on Various Subjects Connected with the Survey of the Coast of the United States," 406–8.

8. Ibid., 406.

9. Monmonier, "Practical and Emblematic Roles of the American Polyconic Projection."

10. Defense Mapping Agency, *Supplement to Department of Defense World Geodetic System 1984 Technical Report: Part I—Methods, Techniques, and Data Used in WGS 84 Development*, ch. 6, p. 7; and Smith, *Introduction to Geodesy: The History and Concepts of Modern Geodesy*, 35. A theoretical surface with zero elevation, the geoid is defined concisely as "the hypothetical extension of mean sea level beneath land surfaces." Defense Mapping Agency, *Supplement*, ch. 6, p. 1.

11. The term "reduction" is related not only to the reduction of comparatively irregular terrain onto a smooth ellipsoid but also to the need to "reduce the probable error." See Silliman, "The Coast Survey of the United States," esp. 235–37; quotation on 235.

12. The larger the denominator, the more closely the ellipse approximates a circle, and the smaller the denominator, the flatter the ellipsoid. In general, ellipticity for "the figure of the earth" is close to $\frac{1}{300}$.

13. Merriman, *Elements of Precise Surveying and Geodesy*, 168.

14. Ibid. Coast and Geodetic Survey scientists were also concerned with measuring and mapping gravity anomalies, especially deflections of the plumb line. Also see Schott, "Recent Contributions to Our Knowledge of the Earth's Shape and Size, by the United States Coast and Geodetic Survey."

15. The last arc was surveyed in 1926. Calculations, which commenced the follow year, were completed in 1932. Dracup, "History of Horizontal Geodetic Control in the United States"; and Schwartz, "North American Datum of 1927."

16. Smith, *Introduction to Geodesy*, 31, 95–97.

17. Within the conterminous Unites States, the shift in position resulting from the new datum was greatest along the west coast but also high in south Florida. See Stem, "User Participation and Impact," 241–42. Also see Wade, "Impact of North American Datum of 1983."

18. For NOAA, converting a thousand charts to a new datum was a huge task. See Stembel, "Horizontal Datum Anomalies on Nautical Charts: A Solution"; and Stembel and Monteith, *Implementation of North American Datum of 1983 into the NOS Nautical Charting Program*.

19. Readily available over the Internet, these announcements became more timely in August 2005, when the Coast Guard increased the frequency of announcements from once a month to once a week.

20. Rizos, "Introducing the Global Positioning System."

21. Bossler, "Datums and Geodetic Systems," esp. 19–21. For the history of WGS 60, an earlier intercontinental datum designed for intercontinental ballistic missiles, see Deborah Jean Warner, "Political Geodesy: The Army, the Air Force, and the World Geodetic System of 1960."

22. For a comparison with earlier electronic navigation systems like LORAN (LOng RAnge Navigation), see Larkin, *Basic Coastal Navigation*, esp. 167–95.

23. National Geospatial-Intelligence Agency, "Using Nautical Charts with Global Positioning System, Edition 3."

24. For an introduction to differential GPS principles, see Ackroyd and Lorimer, *Global Navigation*, 45–68. Also see John Nowatzki and others, "GPS Applications in Crop Production."

25. Authoritative sources differ markedly if not substantially on the precision of GPS for marine navigation. The National Geospatial Intelligence Agency, which provides charts for international waters, asserts that "an accuracy of 2 to 7 meters (same as the military user) *can* be achieved" (emphasis mine). National Geospatial-Intelligence Agency, "Using Nautical Charts with Global Positioning System, Edition 3." The ever-cautious Coast Guard claims that its "DGPS Service provides 10-meter accuracy in all established coverage areas," but in the very next sentence notes that "typically, the positional error of a DGPS position is 1 to 3 meters." U.S. Coast Guard Navigation Center, "DGPS General Information." Garmin International, a prominent vendor, reports "3–5 meters" as "typical differential GPS (DGPS) position accuracy" and "<3 meters" as "typical WAAS position accuracy." Garmin International, "What is WAAS?" Also see Lethan, *GPS Made Easy*, 7, 23–28, 187–89.

26. The federal government first announced horizontal accuracy standards in 1941, with a focus on topographic mapping. Thompson, *Maps for America*, 102–7. Two key Coast and Geodetic Survey publications on chart making from the 1920s and 1930s say nothing about horizontal accuracy. See Deetz, *Cartography*; and Flower, *Rules and Practice Relating to Construction of Nautical Charts*. For discussion of the 1.5-mm standard and related issues, see National Oceanic and Atmospheric Administration, "Chart Accuracy." Until the 1980s, when side-scan sonar began to paint a more complete picture of the seafloor, subsurface hydrography was the primary restraint on accuracy. Key concerns of chart makers were the horizontal separation of sounding lines and a chart's inability to show all available soundings. See, for example, Shalowitz, *Interpretation and Use of Coast and Geodetic Survey Data*, 288–90.

27. National Oceanic and Atmospheric Administration, "Chart Accuracy."

28. Calder, *How to Read a Nautical Chart*, 33–37; and Perugini, "Behind the Accuracy of Electronic Charts." A gateway to individual Web pages describing NOAA's various charting formats and programs is NOAA Office of Coast Survey, "Navigational Charts."

29. Enabnit and Jadkowski, "Electronic Update Service for Digital Charts."

30. Loeper and Shadid, "Digital Update Service and Print-on-Demand."

31. Calder, *How to Read a Nautical Chart*, 37–40.

32. Ibid., 41.

33. Enabnit and Brown, "Electronic Charting at NOAA"; and Gardner,

Ries, and Ware. "Electronic Navigation with Standards and Digital Charts." For discussion of technical standards for electronic nautical charting by the United States and other nations, see Alexander and others, "Future Edition of IHO S-57 (4.0)."

34. The ENC image in the right half of figure 6.7 was generated on a computer screen by SeeMyDEnc, a chart viewer provided without charge by SevenCs, a supplier of Electronic Chart Display and Information Systems (ECDISs). See the firm's Web site at www.sevencs.com.

35. Whereas NOAA scanned shorelines and other natural features for its ENCs from existing images used to produce paper charts, features "critical to the safety of navigation" were traced from more accurate 1:2,400 drawings. See the response to "Is a NOAA ENC a simply a copy of the paper chart?" at NOAA Office of Coast Survey, "Frequently Asked Questions."

36. Wells, "Representing Uncertainty on Charts." For discussion of related research at the University of New Hampshire, see Brennen and others, "Electronic Chart of the Future"; and Plumlee and others, "The CCOM Chart-of-the-Future Project."

37. Enabnit and Jadkowski, "Electronic Update Service for Digital Charts."

38. NOAA, National Ocean Service, *Coast Pilot 1—Atlantic Coast: Eastport to Cape Cod*, 35th ed., 281. Complete, electronic Coast Pilots are available free at http://nauticalcharts.noaa.gov/nsd/cpdownload.htm.

CHAPTER 7

1. In 1884, for instance, Ernest Knorr, a draftsman in Washington, D.C., recommended that hydrographic agencies exchange printing plates, or at least label the intended dimensions of each printed chart so that anyone using it to compile another chart might adjust for paper shrinkage. See Knorr, *A Few Words on International Cooperation in Maritime Hydrography*.

2. Heffernan, "The Politics of the Map in the Early Twentieth Century," 209–10.

3. Crawford, "The Universe of International Science, 1880–1939."

4. Penck, "The Construction of a Map of the World on a Scale of 1:1,000,000," 254.

5. Ibid.

6. Ibid., 259.

7. Hinks, *Maps and Survey*, 81–87.

8. International Map Committee, "International Map of the World."

9. Ibid., 184.

10. Joerg, "Development and State of Progress of the United States Portion of the International Map of the World," 842.

11. Willis, "The International Millionth Map of the World," 131.

12. Ibid.

13. Sheet names and publication dates of the first four sheets were confirmed by information on the maps themselves as well as entries in the *Monthly Catalog of United States Public Documents* for November 1912 (p. 231), April 1915 (p. 571), October 1915 (p. 156), and September 1927 (p. 161).

14. U.S. Department of State, *International Map of the World*. The United

States had dropped out around 1915, when World War I halted the coordination effort.

15. Hinks, *Maps and Survey*, 85–86; quotation on 86. Hinks notes that this opinion, originally prepared for an earlier edition of his book, was part of "a brief statement of the position in August 1931." Ibid., 85.

16. Ibid., 86.

17. Hinks, "The Map on the Scale 1/1,000,000, Compiled at the Royal Geographical Society under the Direction of the General Staff, 1914–1915"; and Hinks, "The 1/Million Map of Europe." The British military's perceived need for the maps was driven by optimistic anticipation of a quick victory. See Heffernan, "Geography, Cartography, and Military Intelligence: the Royal Geographical Society and the First World War," esp. 519–20.

18. Macleod, "The International Map," 446.

19. Ibid.

20. Martin, *The Life and Thought of Isaiah Bowman*, 81–97; and Smith, *American Empire: Roosevelt's Geographer and the Prelude to Globalization*, 113–38.

21. Annual Report of the Council of the American Geographical Society, quoted by AGS historian John K. Wright, in Wright, *Geography in the Making*, 300.

22. Letter from Bowman to Wright, written in 1949 and quoted in Wright, *Geography in the Making*, 301.

23. Platt, "The Map of Hispanic America on the Scale of 1:1,000,000"; Smith, *American Empire*, 92–97; and Wright, *Geography in the Making*, 300–14.

24. Wright, *Geography in the Making*, 308.

25. Quotation from Lord Rennell, president of the Royal Geographical Society, in Rennell, Bowman, and Halifax, "The Geographical Situation of the United States in Relation to World Policies: Discussion," 143.

26. Map sheets and publication dates are listed in United Nations Secretariat, *International Map of the World on the Millionth Scale: Report for 1955*. For discussion of Army Map Service compilation work at 1:250,000 and 1:1,000,000 overseas, see Alexander, "Compiling Medium and Small-scale Maps." Each 1:1,000,000 AMS map listed the corresponding 1:250,000 maps as its principal sources. After completing 1:250,000 coverage nationwide, the AMS turned the series over to the USGS; see Kotch, "The 1:250,000-scale Map Series of the United States."

27. Thompson, *Maps for America*, 32. Similar wording appears on page 127.

28. United Nations Secretariat, "The International Map of the World on the Millionth Scale and the International Co-operation in the Field of Cartography." Sanctioned by the International Geographical Union as well as by the nations involved with the International Map, the formal handover occurred on September 30, 1953.

29. With the cooperation of its allies, the United States military completed the 1:1,000,000-scale World Aeronautical Chart during World War II. See Bahn, "World Aeronautical Charts," 15; and Ristow, *Aviation Cartography*, 32. Coverage available to civilians deteriorated during the Cold War, when the status map issued by the International Civil Aviation Organization showed huge gaps in China and the Soviet Union, where up-to-date aero-

nautical charts had become military secrets. See, for example, International Civil Aviation Organization, *ICAO Aeronautical Chart Catalogue*, July 30, 1965, page 2-3-3.

30. Robinson, "The Future of the International Map," quotation on 24.

31. Ibid., 25.

32. Winchester, "Taking the World's Measure."

33. Scott, and others, *The History of GEBCO 1903–2003*.

34. The International Hydrographic Organization was founded as the International Hydrographic Bureau, the current name of the agency running the IHO's headquarters in Monaco.

35. For a description and background, see Carpine-Lancre, "The Origin and Early History of *La Carte générale bathymétrique des oceans*."

36. Although the 72nd parallels north and south are straight lines on the Mercator map and circles on the adjoining gnomonic projections, along these seams the projections share the same scale, 1:3,060,776, to be precise. See Nichols, "International Co-operation and Co-ordination in Cartography as Viewed by Hydrographers."

37. The first sheet of the more carefully compiled second edition was issued in 1912 and the last in 1930. Ibid., 32–45.

38. For a description and history of the third and fourth editions, see Kerr, "The International Hydrographic Bureau Period."

39. Scott, "A Change of Direction."

40. To help users assess the reliability of isobaths based on soundings that varied markedly in density, the digital atlas included the tracklines of sounding vessels. See Jones, "GEBCO Enters the Digital Era—the GEBCO Digital Atlas."

41. Soluri and Woodson, "World Vector Shoreline."

42. U.S. Department of the Navy, *DONCIO Term Glossary* (online), s.v. "Digital Landmass Blanking System."

43. Mroz and others, "The Defense Mapping Agency (DMA) World Vector Shoreline Plus (WVSPLUS) Revision and Conversion Project."

44. National Oceanic and Atmospheric Administration, Coastal Services Center, "World Vector Shoreline Plus."

45. Determined to rule out an accident at our end, Syracuse University map librarian John Olson verified that NOAA's own intermediate-scale shoreline file, derived directly from NOAA charts and certified for display at 1:70,000, contains Georgetown Island. The original WVS dataset, developed by the Defense Mapping Agency (now the National Geospatial-Intelligence Agency) from its Digital Landmass Blanking Data, does not. The island is also missing from the 1:1,000,000-scale WVS+ library, derived from the original 1:250,000 WVS data.

46. Warita and Nonomura, "The National and Global Activities of the Japanese National Mapping Organization."

47. As of this writing, VMAP1 coverage available to the public included substantially less than half of the world's land area. For an online status map, see National Geospatial-Intelligence Agency. "NGA Raster Roam."

48. Estes and others, "The Imperative for Proactive Cooperative National

Mapping Strategies in the New Millennium"; and Lenczowski, "The Military as Users and Producers of Global Spatial Data."

49. Rhind, "Current Shortcomings of Global Mapping and the Creation of a New Geographical Framework for the World," esp. 301–3.

CHAPTER 8

1. Signed on December 27, 1988, the proclamation was published two weeks later in the *Federal Register*. Reagan, "Proclamation 5928 of December 27, 1988: Territorial Sea of the United States of America." I estimated the size of the addition by first finding the total area of the three-mile territorial sea. In the late 1950s State Department geographer G. Etzel Pearcy had estimated the territorial seas of the conterminous United States and Hawaii as 17,321 and 3,069 square nautical miles, respectively. See Pearcy, "Hawaii's Territorial Sea"; and Pearcy, "Measurement of the U.S. Territorial Sea." I derived a corresponding estimate of 12,972 square nautical miles for Alaska by assuming a similar ratio between that state's 6,640 miles of general coastline and the 10,442 miles of general coastline for the other forty-nine states. I multiplied the total of 33,362 square nautical miles by a conservative 2.5 (rather than 3.0) for the Reagan increment to get 83,405 square nautical miles, which converts to 110,453 square statute miles, the measure customarily used to compare land areas.

2. Reagan, "Proclamation 5928 of December 27, 1988: Territorial Sea of the United States of America."

3. For fuller discussion of the rights associated with the various maritime territories mentioned in this chapter, see Prescott and Schofield, *The Maritime Political Boundaries of the World*.

4. This addition, which may not extend the EEZ more than 350 nautical miles outward from the country's baseline or more than 100 nautical miles beyond the 2,500 meter isobath, must be approved by the UN Commission on the Limits of the Continental Shelf. Ibid., 185–211.

5. Steinberg, *The Social Construction of the Ocean*, 138–50.

6. Reagan, "Proclamation 5030 of March 10, 1983: Exclusive Economic Zone of the United States of America."

7. U.S. Geological Survey, "USEEZ: Boundaries of the Exclusive Economic Zones of the U.S. and Territories."

8. Clinton, "Proclamation 7209 of August 2, 1999: Contiguous Zone of the United States."

9. United Nations, Division of Ocean Affairs and the Law of the Sea, Chronological Lists of Ratifications of, Accessions and Successions to the Convention and the Related Agreements.

10. See Browne, "The U.N. Law of the Sea Convention and the United States: Developments Since October 2003"; and Turner, "Accession to the 1982 Law of the Sea Convention and Ratification of the 1994 Agreement Amending Part XI of the Law of the Sea Convention."

11. Quotation from Article 7 of the UN Convention on the Law of the Sea, online at United Nations, Oceans and Law of the Sea Web site (http://www.un.org/Depts/los/index.htm).

12. See, for example, Smith, "The Maritime Boundaries of the United States."

13. Prescott and Schofield list the twenty-three counties, the lengths of all their straight baselines longer than fifty nautical miles, the date declared, and distance from the coast. See Prescott and Schofield, *The Maritime Political Boundaries of the World*, 654–55.

14. U.S. Department of State, Bureau of Oceans and International Environmental and Scientific Affairs, "Maritime Boundary: Cuba–United States," 2.

15. Ibid., 3.

16. A State Department report on the boundary settlement suggests an ulterior motive for the compromise insofar as American diplomats rarely cooperated with Castro's minions. "Comparing the negotiated boundary line to the original line published by the United States in March 1977," the report notes, "the United States gained a net area of approximately 235 square nautical miles; a slight loss in the eastern part of the boundary area, offset in the central and western portions of the boundary." Ibid., 4. Even so, the Senate declined to ratify the treaty, apparently until late 2003, according to an entry on the State Department's Web site: "Agreement extending the provisional application of the maritime boundary agreement of December 16, 1977. Exchange of notes at Washington and Havana December 16 and 30, 2003; entered into force December 30, 2003." See U.S. Department of State, Bureau of Public Affairs, *Treaties in Force 2005*, Section 1: Bilateral and Other Agreements, 72.

17. Smith, "The Maritime Boundaries of the United States," esp. 402–4.

18. The Senate Committee on Foreign Relations unanimously approved the treaty, which went into the "pending" pile after it was removed from debate to allow further consideration of its terms. See U.S. Congress, Senate Committee on Foreign Relations, *U.S.–Mexico Treaty on Maritime Boundaries*, esp. pp. 2–3.

19. Hedberg's testimony and supporting documents appear in U.S. Congress, House Committee on Foreign Relations, *Three Treaties Establishing Maritime Boundaries between the United States and Mexico, Venezuela, and Cuba*, 28–52; quotation on 31.

20. U.S. Congress, Senate Committee on Foreign Relations, *U.S.–Mexico Treaty on Maritime Boundaries*; quotation on 5.

21. U.S. Department of State, Bureau of Public Affairs, *Treaties in Force 2005*, Section 1: Bilateral and Other Agreements, 201.

22. For additional information consult one of the massive texts on maritime boundaries, such as Klein, *Dispute Settlement in the UN Convention on the Law of the Sea*; or Prescott and Schofield, *The Maritime Political Boundaries of the World*.

23. Quotation from Article 121 of the UN Convention on the Law of the Sea, online at United Nations, Oceans and Law of the Sea Web site (http://www.un.org/Depts/los/index.htm).

24. Ibid., Article 13.

25. Ibid., Article 10, paragraph 6; and Prescott and Schofield, *The Maritime Political Boundaries of the World*, 114–16.

26. U.S. Department of State, Bureau of Oceans and International Envi-

ronmental and Scientific Affairs, "United States Responses to Excessive National Maritime Claims," 8.

27. Ibid., 8–12. Over eighty nations had claimed territorial seas or EEZs that the United States considered "excessive."

28. Ibid., 18.

29. Ibid., 17.

30. Notes in U.S. Department of Defense, Undersecretary of Defense for Policy, *Maritime Claims Reference Manual*, table for Libya, online at http://www.dtic.mil/whs/directives/corres/20051m_062305/Libya.doc (accessed January 12, 2006).

31. Shelley, "Law of the Sea: Delimitation of the Gulf of Maine."

32. Quotation from dissenting opinion of Judge Gros; see International Court of Justice, "Case Concerning Delimitation of the Maritime Boundary in the Gulf of Maine Area," 360.

33. Unnamed United States counsel quoted in separate opinion of Judge Schwebel; see ibid., 353.

34. The United States might have fared better had the World Court judges found the colors on its maps more appealing. Joseph Wiedel, who prepared the maps used by the American legal team, learned the nationalities of panel members, consulted studies of cultural effects on color preference, and proposed a visually attractive design that underscored the U.S. position. He was overruled by a State Department official who favored brilliant reds and blues, considered harsh or threatening in some cultures. Joseph W. Wiedel, telephone interview, February 2, 2007.

35. De Vorsey, "Florida's Seaward Boundary: A Problem in Applied Historical Geography"; and Reed, *The Development of International Maritime Boundary Principles through United States Practice*, 23–32.

36. For an authoritative analysis of the Dinkum Sands controversy, see Reed, *The Development of International Maritime Boundary Principles through United States Practice*, 133–40, 204–10.

37. U.S. Supreme Court, *United States v. Alaska*, 29–30.

38. Ibid., 32. Alaska also lost its more significant claim to submerged lands within the Arctic National Wildlife Refuge and the National Petroleum Reserve–Alaska.

39. Reed, *The Development of International Maritime Boundary Principles through United States Practice*, 135–36.

40. The NOAA Corps Collection in NOAA's online photo library includes a picture of marker taken by Ensign Nygren in summer 1949: http://www.photolib.noaa.gov/htmls/corp1068.htm.

41. The Coastline Committee, established in 1970 as the Ad Hoc Committee on the Delimitation of the United States Coastline, is also known as the Baseline Committee. See Harrington, "Maritime Boundaries on National Ocean Service Nautical Charts," 9n1.

CHAPTER 9

1. We'd also be living in the early twelfth "grossury." In a dozenal system, with counting based on twelve, not ten, the number "100" would mean 144 in

our base-ten counting system, and twelve "dozades" (each twelve years long) would make up a grossury, with 144 decimal years.

2. For a definition and concise history of the term, see Jenkins, "Social Construction."

3. For examples, see Beatley, Brower, and Schwab, *An Introduction to Coastal Zone Management*, 124–26; and Mercado, "On the Use of NOAA's Storm Surge Model, SLOSH, in Managing Coastal Hazards: The Experience in Puerto Rico."

4. Regulations and policy considerations are far more involved than indicated here, and subject to controversy and incremental change. For an overview and critical assessment of the National Flood Insurance Program, see Burby, "Flood Insurance and Floodplain Management: the U.S. Experience"; and Crowell, Hirsch, and Hayes, "Improving FEMA's Coastal Risk Assessment through the National Flood Insurance Program: An Historical Overview."

5. For an overview of the development and use of flood insurance maps, see Monmonier, *Cartographies of Danger*, 107–26.

6. Gene Stakhiv, U.S. Army Corps of Engineers, interviewed July 28, 1994, and quoted in Monmonier, *Cartographies of Danger*, 122.

7. Interpolation refers to the estimation of intermediate values from known measurements or probability-based estimates. Intermediate values are calculated as weighted averages, with the flood elevation estimated for the closer gauging station accorded a higher weight than the one farther up- or downstream. Weights are based on distance along the river, rather than overland distance.

8. Wind speeds and surge heights are from National Oceanic and Atmospheric Administration, National Weather Service, National Hurricane Center, "The Saffir-Simpson Hurricane Scale."

9. For a detailed description of SLOSH modeling, including the partial differential equations of motion used to represent surface winds, see Jelesnianski, Chen, and Shaffer, *SLOSH: Sea, Lake, and Overland Surges from Hurricanes*. For an example of an earlier numerical approach combining a regular grid offshore with an irregular grid for coastal embayments, see Dendrou, Moore, and Myers, "Application of Storm Surge Modeling to Coastal Flood Rate Determinations." For a comparison of SLOSH with other models, see Massey and others, "History of Coastal Inundation Models."

10. Kleinberg, *Black Cloud: The Great Florida Hurricane of 1928*; and Zhang, Xiao, and Leatherman, *Storm Surge Simulation for Lake Okeechobee*.

11. National Oceanic and Atmospheric Administration, National Weather Service, Evaluation Branch, "Hurricane Storm Surge Forecasting."

12. Significant surge can also occur to the left of the storm track behind the eye.

13. Crawford, "Hurricane Surge Potentials over Southeast Louisiana as Revealed by a Storm-Surge Forecast Model: A Preliminary Study," 423–24. A shorter time interval can make the model more computationally intensive, as would a finer mesh, but problematic data and the inherent uncertainty of numerical modeling might not justify this effort.

14. See, for example, Stone, Zhang, and Sheremet, "The Role of Barrier

Islands, Muddy Shelf and Reefs in Mitigating the Wave Field along Coastal Louisiana."

15. Jelesnianski, Chen, and Shaffer, *SLOSH: Sea, Lake, and Overland Surges from Hurricanes*, 46–49, 51. "Surge envelope" refers to the mass of water between the sea's normal tidal level and its higher, surge-induced level.

16. Feyen and others, "Development of a Continuous Bathymetric/Topographic Unstructured Coastal Flooding Model to Study Sea Level Rise in North Carolina"; and Shen and others, "Improved Prediction of Storm Surge Inundation with a High-Resolution Unstructured Grid Model."

17. Florida International University, International Hurricane Research Center, Laboratory for Coastal Research, "Storm Surge Model Evalation."

18. Florida International University, International Hurricane Research Center, Laboratory for Coastal Research, *Windstorm Simulation and Modeling: Executive Summary*, September 2004, Miami, FL; and Houston and others, "Comparisons of HRD and SLOSH Surface Wind Fields in Hurricanes: Implications for Storm Surge Modeling."

19. U.S. Army Corps of Engineers, Philadelphia District, "Hazards and Vulnerability Data."

20. Stephen Leatherman, telephone interview, February 8, 2007.

21. For example, a concise account of the use of SLOSH modeling in Puerto Rico devotes only three pages to flood insurance studies. See Mercado, "On the Use of NOAA's Storm Surge Model, SLOSH, in Managing Coastal Hazards—The Experience in Puerto Rico."

22. The sample Town of Floodport report is part of Appendix C in Federal Emergency Management Agency, *Guidelines and Specifications for Flood Map Production Coordination Contractors: Appendices and References*.

23. Ibid., C-24. For a concise history of coastal flood mapping, including technical and policy issues, see Bellomo, Pajak, and Sparks, "Coastal Flood Hazards and the National Flood Insurance Program"; and Crowell, Hirsch, and Hayes, "Improving FEMA's Coastal Risk Assessment through the National Flood Insurance Program: An Historical Overview."

24. Federal Emergency Management Agency, *Guidelines and Specifications for Flood Map Production Coordination Contractors: Appendices and References*, G-1. For a list of improved models used in FEMA's more recent flood studies, see Federal Emergency Management Agency, *Numerical Models Meeting the Minimum Requirements of the NFIP*.

25. National Research Council, Committee on Coastal Flooding from Hurricanes, *Evaluation of the FEMA Model for Estimating Potential Coastal Flooding from Hurricanes and Its Application to Lee County, Florida*, 2.

26. Ibid., 5.

27. Ibid., 118.

28. Ibid., 119.

29. Ibid., 40.

30. For the environmentalist perspective, see Beatley, Brower, and Schwab, *An Introduction to Coastal Zone Management*, 124–26. For insights into cost issues that frighten fiscal conservatives, see Randall, "Coastal Development Run Amuck: A Policy of Retreat May Be the Only Hope."

31. Quotes in Nassar, "Storms Don't Deter N.C. Coastal Residents."

32. U.S. General Accounting Office, *National Flood Insurance: Marginal Impact on Flood Plain Development, Administrative Improvements Needed*, 12.

33. U.S. General Accounting Office, *National Flood Insurance Program: Actions to Address Repetitive Loss Properties*, 7. For an overview of FEMA's map modernization project, see Lowe, "The Federal Emergency Management Agency's Multi-Hazard Flood Map Modernization and *The National Map*."

34. Crowell, Leiken, and Buckley, "Evaluation of Coastal Erosion Hazards Study: An Overview"; and Leatherman and others, *Evaluation of Erosion Hazards: Summary*.

35. King, "Federal Flood Insurance: The Repetitive Loss Problem," 20.

36. Ibid., 22.

37. Lee County, Florida, GIS Department, *Lee County Flood Zones* (map).

38. Zones A, AE, and V are not the only areas of 100-year flooding. I omitted zones B, C, and X, defined in technically complex prose in the map key. Zone B comprises "areas between limits of the 100-year flood and 500-year flood; or certain areas subject to 100-year flooding with average depths less than (1) foot or where the contribution drainage area is less than one square mile; or areas protected by levees from the base flood." Zone X, parts of which are also subject to 100-year flooding, is defined as the "Areas of 500-year flood; areas of 100-year flood with average depths of less than 1 foot or with drainage areas less than 1 square mile; and areas protected by levees from [the] 100-year flood." By contrast, Zone C is merely labeled "Areas of minimal flooding," which could be more frequent than 100 years. About 5 percent of the county, higher or better drained, is classified as "other." Ibid.

39. Lee County, Florida, Department of Community Development, *The Lee Plan 2004 Codification, as Amended through December 2004*, VII-3 to VII-4, XII-3.

40. Florida Department of Community Affairs, *Coastal High Hazard Study Committee Final Report*, 7. In 2007 a National Research Council committee recommended use of lidar to update elevation data nationwide, with riverine as well as coastal flood mapping as a crucial application. See National Research Council, Committee on Floodplain Mapping Techniques, *Base Map Inputs for Floodplain Mapping*.

41. The Web map's key reports a range of elevations above mean sea for each surge category. Because storm surge is not solely an effect of elevation, the ranges overlap: tropical storm (4.1 to 5.6 feet), category 1 (4.4 to 7.4 feet), category 2 (7.9 to 12.4 feet), category 3 (11.2 to 19.5 feet), and categories 4, 5, and 6 (16.5 to 26.7 feet). Lee County (Florida) Emergency Management, *Lee County Elevation and Evacuation Map*.

42. For examples of computer and mathematical modeling of tsunami inundations, see Bernard, "Reducing Tsunami Hazards along U.S. Coastlines"; Bryant, *Tsunami: The Underrated Hazard*, esp. 25–59, 265–93; and Koike and others, "Tsunami Run-up Heights of the 2004 Off the Kii Peninsula Earthquakes." For a frank appraisal of the accuracy of inundation maps, see U.S. Government Accountability Office, *U.S. Tsunami Preparedness: Federal and State Partners Collaborate to Help Communities Reduce Potential Impacts, but Significant Challenges Remain*.

43. Fischetti, "Drowning New Orleans," 78.

44. Bourne, "The Big Uneasy," 92.

45. A February 2006 evaluation of emergency operations put the death toll at 1,330 but noted that 2,096 more were still missing. By late December 2005, roughly 500,000 of the 1.1 million persons evacuated had not returned. See Townsend, *The Federal Response to Hurricane Katrina: Lessons Learned*, 8–9. For an insightful appraisal of plans to rebuild the city, see Kates and others, "Reconstruction of New Orleans after Hurricane Katrina: A Research Perspective."

CHAPTER 10

1. According to a recent report by the Intergovernmental Panel on Climate Change (IPCC), sea level is expected to increase worldwide between 18 and 59 centimeters over the period 1980 to 2100. See Intergovernmental Panel on Climate Change, *Climate Change 2007: The Physical Science Basis—Summary for Policymakers*. An earlier prediction suggested a wider range of outcomes. See, for example, Barnett and Adger, "Climate Dangers and Atoll Countries," 323.

2. Greve, "On the Response of the Greenland Ice Sheet to Greenhouse Climate Change."

3. For a history and critique of the Davisian erosion cycle, see Chorley, Beckinsale, and Dunn, *The Life and Work of William Morris Davis*, 160–233.

4. Johnson, *Shore Processes and Shoreline Development*, 275, 283.

5. Ibid., 190. Johnson did not ignore measurements that supported his arguments. In an article published two years before *Shore Processes* but not mentioned therein, he cited several geodetic studies that suggested the Atlantic shoreline was comparatively stable—and thus consistent with his notions of emerging and stable coastlines. See Johnson, "Is the Atlantic Coast Sinking?"

6. Marmer, "Sea Level along the Atlantic Coast of the United States and Its Fluctuations," 448.

7. Although the literature abounds with an appropriate diversity of rates, the 18- and 12-centimeter averages discussed with no apparent objection at a 2001 National Academy of Sciences roundtable for climate scientists and government policy makers seems an acceptably concise and defensible summary. See Leatherman and Kershaw, *Sea Level Rise and Coastal Disasters: Summary of a Forum, October 25, 2001*, 2.

8. See National Research Council, Committee on Options for Preserving Cape Hatteras Lighthouse, *Saving Cape Hatteras Lighthouse from the Sea: Options and Policy Implications*; and Riddle, "Crawling, a Lighthouse Flees the Battering Sea."

9. See Leatherman and Kershaw, *Sea Level Rise and Coastal Disasters*, 7. East Carolina University geologist Stanley Riggs has been particularly outspoken. According to Leatherman and Kershaw, "Riggs speculated that in 200 years almost no North Carolina barrier islands would be left." See also Pilkey and others, *The North Carolina Shore and Its Barrier Islands: Restless Ribbons of Sand*.

10. For a concise chronology of coastal erosion and efforts to protect the

lighthouse, see Pilkey and others, *The North Carolina Shore and Its Barrier Islands: Restless Ribbons of Sand*, 1–9.

11. National Research Council, Committee on Options for Preserving Cape Hatteras Lighthouse, *Saving Cape Hatteras Lighthouse from the Sea: Options and Policy Implications*, esp. summary on 5–6.

12. Schneider and Chen, "Carbon Dioxide Warming and Coastline Flooding: Physical Factors and Climatic Impact," 107.

13. For the Washington, D.C., map, see ibid., fig. 2, 120–21.

14. Ibid., 117.

15. Ibid., 140.

16. Kopec, "Global Climate Change and the Impact of a Maximum Sea Level on Coastal Settlement," 542.

17. Ibid., 549.

18. For an examination of diversity and uncertainty among climatologists and coastal scientists, see Posner, *Catastrophe: Risk and Response*, 43–58; and Titus and Narayanan, "The Risk of Sea Level Rise."

19. Titus and Richman. "Maps of Lands Vulnerable to Sea Level Rise: Modeled Elevations along the U.S. Atlantic and Gulf Coasts," 210–11.

20. National Research Council, Committee on Engineering Implications of Changes in Relative Mean Sea Level, *Responding to Changes in Sea Level: Engineering Implications*, 53–61.

21. For applications of the Bruun Rule, see Leatherman, "Social and Economic Costs of Sea Level Rise"; and Psuty and Ofiara, *Coastal Hazard Management: Lessons and Future Directions from New Jersey*, 84–104. For a vigorous critique, see Cooper and Pilkey, "Sea-Level Rise and Shoreline Retreat: Time to Abandon the Bruun Rule."

22. National Research Council, Committee on Coastal Erosion Zone Management, *Managing Coastal Erosion*, 122–29. After carefully registering multiple historic shorelines onto a contemporary map, scientists calculate rates of shoreline displacement along profiles perpendicular to the coast and use these rates to estimate a shoreline's position ten, twenty-five, or even fifty years hence. Map overlay also affords a dramatic description in the variability of shoreline erosion or accretion as well as the effects of longshore currents, as at Cape Hatteras, where the coastline's southward shift was more pronounced than its inland retreat.

23. See Titus, "Does the U.S. Government Realize That the Sea Is Rising? How to Restructure Federal Programs So That Wetlands and Beaches Survive," esp. 723–34.

24. For discussion of the rationale for and legal implications of rolling easements, see Titus, "Rising Seas, Coastal Erosion, and the Takings Clause: How to Save Wetlands and Beaches without Hurting Property Owners," esp. 1313–18. For a wider discussion of the social, political, and scientific issues surrounding the loss of beaches and coastal wetlands, see Dean, *Against the Tide*.

25. Titus and Richman, "Maps of Lands Vulnerable to Sea Level Rise: Modeled Elevations along the U.S. Atlantic and Gulf Coasts," quotation on 226.

26. U.S. Environmental Protection Agency, Office of Air and Radiation, Maps of Lands Vulnerable to Sea Level Rise.

27. U.S. Environmental Protection Agency, Office of Air and Radiation, Forthcoming Products.

28. Titus, "Maps That Depict the Business-As-Usual Response to Sea Level Rise in the Decentralized United States of America."

29. Ibid.

30. Image of chart 1220 is from NOAA's online collection of historic nautical charts. See National Oceanic and Atmospheric Administration, Office of Coast Survey, "Historical Maps and Charts—Background."

31. For a concise overview of newfound fears, see Kerr, "A Worrying Trend of Less Ice, Higher Seas." Also see National Research Council, Committee on Abrupt Climate Change, *Abrupt Climate Change: Inevitable Surprises*.

32. Parizek and Alley, "Implications of Increased Greenland Surface Melt under Global-Warming Scenarios: Ice-Sheet Simulations," quotation on 1013.

33. Byron R. Parizek, e-mail communication, March 24, 2006; also see Bellassen and Chameides, *High Water Blues: The Climate Science behind Sea Level Rise and Its Policy Implications*, 10–11.

CHAPTER 11

1. Curated by Alice Hudson, the exhibit "Treasured Maps: Celebrating the Lionel Pincus and Princess Firyal Map Division" was on display from September 9, 2005 through April 9, 2006.

2. Campbell, "Portolan Charts from the Late Thirteenth Century to 1500."

3. Verner, "Copperplate Engraving."

4. Crane, *Mercator: The Man Who Mapped the Planet*, 111–12; quotation on 111.

5. Bagrow, *History of Cartography*, 64; and Campbell, "Portolan Charts," 377.

6. Whiting, "Reports of Assistant H. L. Whiting on Topographical Contour, Hydrographic Details, and Reduction, on Photography, and on the Scale of Shades Suitable for Complete Maps," 219.

7. Ibid., 220–21.

8. Calder, *How to Read a Nautical Chart*, esp. 132–39.

9. National Oceanic and Atmospheric Administration, Office of Coast Survey, *Chart No. 1: Nautical Chart Symbols, Abbreviations, and Terms*, Chapter K, "Rocks, Wrecks, Obstructions," 43–46.

10. Calder, *How to Read a Nautical Chart*, 124.

11. Writing decades after hachures became passé, Coast and Geodetic Survey mapping expert Charles Deetz cited several reasons for their disappearance. "Under certain conditions they are very expressive but are of little use in supplying degree of slope or actual height of ground above high-water or sea-level. In themselves, hachures with their shaded effects frequently do not distinguish readily that which is hill from valley, and interpretation must be based on other details such as rivers or heights at different places." See Deetz, *Cartography*, 70.

12. Shalowitz, *Interpretation and Use of Coast and Geodetic Survey Data*, 192–210, 258–61, 282–85, 683–704.

13. The movement away from dense soundings culminated in the 1939

adoption of a chart style based on echo soundings and allowing a fuller use of comparatively reliable depth contours. Ibid., 282–85.

14. Ibid., 329.

15. A cursory look at samples in my university's map collection found a few examples of 1930s maps with a continuous blue tint for large bodies of water. Even so, for most areas the older style typically persisted until the 1950s and 1960s, when 7.5-minute quadrangle maps at 1:24,000 replaced 15-minute mapping at 1:62,500.

16. My count of thirty-eight does not include the map's noticeably thicker blue shoreline.

17. Allix, "The Geography of Fairs: Illustrated by Old-World Examples."

18. According to one postwar cartographic textbook, "when skillfully done, a casual placing of the dots may give a more even impression than would a strict alignment of the dots, but it needs considerable practice to achieve this." Monkhouse and Wilkinson, *Maps and Diagrams: Their Compilation and Construction*, 37.

19. Light from the top left, which is more common, is less satisfactory for this map's coastlines.

20. Citizens' Association of New York, Council of Hygiene and Public Health, *Report of the Council of Hygiene and Public Health of the Citizens' Association of New York upon the Sanitary Condition of the City*, map inserted between pages xx and xxi.

21. For examples, see Costello, "Beach Access: Where Do You Draw the Line in the Sand"; and Davidson and Entrikin, "The Los Angeles Coast as a Public Place."

22. For fuller information about access issues too detailed for a small-scale national map, see Titus, "Rising Seas, Coastal Erosion, and the Takings Clause: How to Save Wetlands and Beaches without Hurting Property Owners," esp. 1364–68.

CHAPTER 12

1. National Research Council, Committee on Planning for Catastrophe, *Successful Response Starts with a Map: Improving Geospatial Support for Disaster Management*, 32–34.

2. Gore, *An Inconvenient Truth: The Planetary Emergency of Global Warming and What We Can Do About It*, 198–209.

3. Ibid., 196–97.

BIBLIOGRAPHY

Ackroyd, Neil, and Robert Lorimer. *Global Navigation: A GPS User's Guide*. 2nd ed. London: Lloyd's of London Press, 1994.

Alba, María del Rosario Falcó y Osorio, Duquesa de. *Nuevos Autógrafos de Cristóbal Colón y Relaciones de Ultramar*. Madrid, 1902.

Alexander, Lee, and others. "Future Edition of IHO S-57 (4.0)." *International Hydrographic Review*, n.s., 6 (April 2005): 66–72.

Alexander, Paul. "Compiling Medium and Small-scale Maps." *Military Engineer* 39 (1947): 302.

Allix, André. "The Geography of Fairs: Illustrated by Old-World Examples." *Geographical Review* 12 (1922): 532–69.

Anderson, Lauren Bayne. "A Quest to Claim America's Birth Certificate." *St. Petersburg Times*, December 22, 2004.

Ascensión, Antonio de la. "A Brief Report of the Discovery in the South Sea." In *Spanish Exploration in the Southwest, 1542–1706*, edited by Herbert Eugene Bolton, 104–34. New York: C. Scribner's Sons, 1916.

Bagrow, Leo. *History of Cartography*. 2nd ed., rev. and enlarged by R. A. Skelton. Chicago: Precedent Publishing, 1985.

Bahn, Catherine I. "World Aeronautical Charts." *Bulletin of the Geography and Map Division, Special Libraries Association*, no. 29 (October 1957): 15–18.

Bargar, B. D. "Samuel Eliot Morison." In *Dictionary of Literary Biography*, vol. 17, *Twentieth-Century American Historians*, edited by Clyde N. Wilson, 296–307. Detroit: Gale Research Co., 1983.

Barnett, Jon, and W. Neil Adger. "Climate Dangers and Atoll Countries." *Climatic Change* 61 (2003): 321–37.

Bay County, Florida. Leisure Services Division. *Beach Access Points Map*. http://www.co.bay.fl.us/community/beach_map.pdf (accessed June 22, 2006).

Beatley, Timothy, David J. Brower, and Anna K. Schwab. *An Introduction to Coastal Zone Management*. 2nd ed. Washington, D.C.: Island Press, 2002.

Bellassen, Valentin, and Bill Chameides. *High Water Blues: The Climate Science*

Behind Sea Level Rise and Its Policy Implications. Updated version. New York: Environmental Defense, 2005. http://www.environmentaldefense.org/documents/4846_HighWaterBlues_Update_2005.pdf.

Bellomo, Doug, Mary Jean Pajak, and Jerry Sparks. "Coastal Flood Hazards and the National Flood Insurance Program." *Journal of Coastal Research,* special issue 28 (Spring 1999): 21–26.

Bencker, Henri. "The Development of Maritime Hydrography and Methods of Navigation." *Hydrographic Review* 21 (1944): 117–29.

Bernard, E. N. "Reducing Tsunami Hazards along U.S. Coastlines." In *Perspectives on Tsunami Hazard Reduction,* edited by Gerald Hebenstreit, 189–203. Dordrecht, The Netherlands: Kluwer Academic Publishers, 1977.

Blachut, Teodor J., and Rudolf Burkhardt. *Historical Development of Photogrammetric Methods and Instruments.* Falls Church, Va.: American Society for Photogrammetry and Remote Sensing, 1989.

Blockade Board. "Second Report of Conference for the Consideration of Measures for Effectually Blockading the South Atlantic Coast." Washington, D.C., July 16, 1861. In U.S. Naval War Records Office, *Official Records of the Union and Confederate Navies in the War of the Rebellion,* series I, vol. 12, 198–201. Washington, D.C.: Government Printing Office, 1901.

Boggs, S. W., and others. "The International Millionth Map of the World." *Hydrographic Review* 6 (November 1929): 181–85.

Bossler, John D. "Datums and Geodetic Systems." In *Manual of Geospatial Science and Technology,* edited by John D. Bossler, 16–26. New York: Taylor and Francis, 2002.

Bourne, Joel K., Jr. "The Big Uneasy." *National Geographic* 206 (October 2004): 88–105.

Boutelle, C. O. "What Has the Coast Survey Done for Science?" *Science* 6 (1885): 558–62.

Brennen, Richard, and others. "Electronic Chart of the Future: The Hampton Roads Demonstration Project." Paper presented at the 2003 U.S. Hydrographic Conference, Biloxi, Miss., March 25–27, 2003. Online at http://www.thsoa.org/hy03/11_3.pdf.

Briggs, Henry. "A Treatise of the Northwest Passage" In *A Declaration of the State of the Colony and Affaires in Virginia . . . ,* by Edward Waterhouse, 45–50. London: Robert Mylbourne, 1622.

Brooks, H. R. "Use of Aerial Photographs for Revision of Land Information on Nautical Charts." *International Hydrographic Review* 32 (November 1955): 129–33.

Browne, Marjorie Ann. "The U.N. Law of the Sea Convention and the United States: Developments Since October 2003." Congressional Research Service Report RS21890, June 3, 2005. Online at http://www.cnie.org/NLE/CRS/detail.cfm.

Bryant, Edward. *Tsunami: The Underrated Hazard.* New York: Cambridge University Press, 2001.

Burby, Raymond J. "Flood Insurance and Floodplain Management: the U.S. Experience." *Environmental Hazards* 3 (2001): 111–22.

Burden, Philip D. *The Mapping of North America: A List of Printed Maps, 1511–1670.* Rickmansworth, Herts., England: Raleigh Publications, 1996.

Cajori, Florian. *The Chequered Career of Ferdinand Rudolph Hassler.* Boston: Christopher Publishing House, 1929; New York: Arno Press, 1980.

Calder, Nigel. *How to Read a Nautical Chart: A Complete Guide to the Symbols, Abbreviations, and Data Displayed on Nautical Charts.* Camden, Maine: International Marine; New York: McGraw-Hill, 2003.

Campbell, Tony. "Portolan Charts from the Late Thirteenth Century to 1500." In *The History of Cartography,* vol. 1, *Cartography in Prehistoric, Ancient, and Medieval Europe and the Mediterranean,* edited by J. B. Harley and David Woodward, 371–463. Chicago: University of Chicago Press, 1987.

Carpine-Lancre, Jacqueline. "The Origin and Early History of *La Carte générale bathymétrique des oceans.*" In *The History of GEBCO, 1903–2003,* edited by Desmond Scott and others, 15–51. Lemmer, The Netherlands: Geomatics Information and Trading Center, 2003.

Cartwright, David Edgar. *Tides: A Scientific History.* Cambridge: Cambridge University Press, 1999.

Chapman, Charles Frederic. *Piloting, Seamanship, and Small Boat Handling.* 51st ed. New York: Motor Boating and Sailing, 1975.

Chorley, Richard J., Robert P. Beckinsale, and Antony J. Dunn. *The Life and Work of William Morris Davis.* Vol. 2 of *The History of the Study of Landforms or the Development of Geomorphology.* London: Methuen, 1973.

Citizens' Association of New York. Council of Hygiene and Public Health. *Report of the Council of Hygiene and Public Health of the Citizens' Association of New York upon the Sanitary Condition of the City.* New York: D. Appleton and Co., 1865.

Clancy, Edward P. *The Tides: Pulse of the Earth.* Garden City, N.Y.: Doubleday and Co., 1968.

Clinton, William J. (President of the United States). "Proclamation 7209 of August 2, 1999: Contiguous Zone of the United States." *Federal Register* 64 (September 8, 1999): 48701–2.

Cloud, John. "The 200th Anniversary of the Survey of the Coast." *Prologue: the Journal of the National Archives* 39 (Spring 2007): 24–33.

Colby, Jeffrey D., Karen A. Mulcahy, and Yong Wang. "Modeling Flooding Extent from Hurricane Floyd in the Coastal Plains of North Carolina." *Environmental Hazards* 2 (2000): 157–68.

Collier, Peter. "The Impact on Topographic Mapping of Developments in Land and Air Survey: 1900–1939." *Cartography and Geographic Information Science* 29 (2002): 155–74.

Collinder, Per. *A History of Marine Navigation.* Translated by Maurice Michael. London: B. T. Batsford, 1954.

Conway, T. M. "Origins of the Terms 'Spring' and 'Neap' Tides." *Mariner's Mirror* 86 (2000): 475–76.

Cooper, J. Andrew G., and Orrin H. Pilkey. "Sea-Level Rise and Shoreline Retreat: Time to Abandon the Brunn Rule." *Global and Planetary Change* 43 (2004): 157–71.

Costello, Jane. "Beach Access: Where Do You Draw the Line in the Sand?" *New York Times*, January 21, 2005.

Cotter, Charles H. *The Complete Nautical Astronomer.* New York: American Elsevier, 1969.

Cox, R. C. "The Development of Survey Instrumentation, 1780–1980." *Survey Review* 28 (1986): 234–55, 283–303.

Crane, Nicholas. *Mercator: The Man Who Mapped the Planet*. New York: Henry Holt, 2003.

Crawford, Elisabeth. "The Universe of International Science, 1880–1939." In *Solomon's House Revisited: The Organization and Institutionalization of Science,* edited by Tore Frängsmyr, 251–69. Canton, Mass.: Science History Publications, 1990.

Crawford, Kenneth C. "Hurricane Surge Potentials over Southeast Louisiana as Revealed by a Storm-Surge Forecast Model: A Preliminary Study." *Bulletin of the American Meteorological Society* 60 (1979): 422–29.

Crowell, Mark, Emily Hirsch, and Tom L. Hayes. "Improving FEMA's Coastal Risk Assessment through the National Flood Insurance Program: An Historical Overview." *Marine Technology Society Journal* 41 (Spring 2007): 18–27.

Crowell, Mark, Howard Leiken, and Michael K. Buckley. "Evaluation of Coastal Erosion Hazards Study: An Overview." *Journal of Coastal Research,* special issue 28 (Spring 1999): 2–9.

Davidson, Ronald A., and J. Nicholas Entrikin. "The Los Angeles Coast as a Public Place." *Geographical Review* 95 (2005): 578–93.

Dean, Cornelia. *Against the Tide: The Battle for America's Beaches.* New York: Columbia University Press, 1999.

Deetz, Charles H. *Cartography.* U.S. Coast and Geodetic Survey Special Publication No. 205. Washington, D.C.: Government Printing Office, 1936.

Deetz, Charles H., and Oscar S. Adams. *Elements of Map Projection with Applications to Map and Chart Construction,* 4th ed., U.S. Coast and Geodetic Survey Special Publication No. 68. Washington, D.C.: Government Printing Office, 1934.

Defense Mapping Agency. *Supplement to Department of Defense World Geodetic System 1984 Technical Report: Part I—Methods, Techniques, and Data Used in WGS 84 Development.* Technical Report No. 8350.2-A. Washington, D.C., 1987. Online at http://earth-info.nga.mil/GandG/publications/historic/historic.html.

Dendrou, Stergios A., Charles I. Moore, and Vance A. Myers. "Application of Storm Surge Modeling to Coastal Flood Rate Determinations." *Marine Technology Society Journal* 19, no. 2 (1985): 42–50.

De Vorsey, Louis, Jr. "Florida's Seaward Boundary: A Problem in Applied Historical Geography." *Professional Geographer* 25 (1973): 214–20.

Dracup, Joseph F. "History of Horizontal Geodetic Control in the United States." In *North American Datum of 1983,* edited by Charles R. Schwartz, 13–20. NOS Professional Paper NOS 2. Rockville, Md.: National Geodetic Survey, 1989.

Edney, Mathew H. "Cartography without `Progress': Reinterpreting the Nature and Historical Development of Mapmaking." *Cartographica* (Summer/Autumn 1993): 54–68.

Enabnit, David B., and Michael B. Brown. "Electronic Charting at NOAA." *Sea Technology* 40 (July 1999): 39–41.

Enabnit, David B., and Mark A. Jadkowski. "Electronic Update Service for Digital Charts." NOAA Office of Coast Survey. April 13, 2003. http://nauticalcharts.noaa.gov/ocs/rnc/update.htm.

Espey, Michael. "Using Commercial Satellite Imagery and GIS to Update NOAA ENCs." Paper presented at annual ESRI International User Conference, San Diego, Calif., July 25–29, 2005.

Estes, John E., and others. "The Imperative for Proactive Cooperative National Mapping Strategies in the New Millennium." In *2001 IEEE Geoscience and Remote Sensing Symposium*, 3:1478–80.

Federal Emergency Management Agency. *Guidelines and Specifications for Flood Map Production Coordination Contractors: Appendices and References*. Final draft, February 17, 1999, http://www.fema.gov/pdf/fhm/frm_spf2.pdf.

———. "Multi-Year Flood Hazard Identification Plan." Version 1.0, November 2004. http://www.fema.gov/plan/prevent/fhm/mh_main.shtm.

———. *Numerical Models Meeting the Minimum Requirements of the NFIP*. http://www.fema.gov/plan/prevent/fhm/en_coast.shtm (accessed February 2, 2007).

Feyen, Jesse, and others. "Development of a Continuous Bathymetric/Topographic Unstructured Coastal Flooding Model to Study Sea Level Rise in North Carolina." *Estuarine and Coastal Modeling—Proceedings of the Ninth International Conference, Charleston, SC, October 31–September 2, 2005*, edited by Malcolm Spaulding, 338–57. Reston, Va.: American Society of Civil Engineers, 2006.

Fischer, Joseph, and Franz von Wieser. Introduction to *The Cosmographiae Introductio of Martin Waldseemüller in Facsimile*, edited by Charles George Herbermann, 6–22. New York: United States Catholic Historical Society, 1907.

Fischetti, Mark. "Drowning New Orleans." *Scientific American* 285 (October 2001): 76–85.

Flood, Martin. "Laser Altimetry: From Science to Commercial Lidar Mapping." *Photogrammetric Engineering and Remote Sensing* 67 (2001): 1209–17.

———. "Product Definitions and Guidelines for Use in Specifying Lidar Deliverables." *Photogrammetric Engineering and Remote Sensing* 68 (2002): 1230–34.

Florida Department of Community Affairs. *Coastal High Hazard Study Committee Final Report*. February 1, 2006. http://www.dca.state.fl.us/fdcp/dcp/chhsc/final2-1-06.pdf.

Florida International University. International Hurricane Research Center. Laboratory for Coastal Research. "Airborne Laser Technology." http://www.ihrc.fiu.edu/lcr/research/airborne_laser_mapping/index.htm.

———. "Storm Surge Model Evaluation." http://www.ihc.fiu.edu/lcr/research/windstorm_simulation/storm_surge_model_evaluation.htm (accessed December 5, 2006).

———. *Windstorm Simulation and Modeling: Executive Summary*. September 2004. Miami, FL.

Flower, George L. *Rules and Practice Relating to Construction of Nautical Charts.* U.S. Coast and Geodetic Survey Special Publication No. 66. Washington, D.C.: Government Printing Office, 1923.

Floyd, Richard P. "National Ocean Service Shoreline Mapping Program." Internal National Geodetic Survey document. ca. 1995. http://www.ngs.noaa.gov/PUBS_LIB/shore_map.html.

Forrest, David, Alastair Pearson, and Peter Collier. "The Representation of Topographic Information on Maps—The Coastal Environment." *Cartographic Journal* 34 (1997): 77–85.

Frey, Jennifer. "Donor's Gift Is New-World Class: Library of Congress to Receive 4,000 Artifacts of Early Americas." *Washington Post,* April 7, 2004.

Gardner, James, Kathryn Ries, and Colin Ware. "Electronic Navigation with Standards and Digital Charts." *Sea Technology* 45 (March 2004): 10–14.

Garmin International. "What Is WAAS?" http://www.garmin.com/aboutGPS/waas.html.

Geary, Edmund L. "Coastal Hydrography." *Photogrammetric Engineering* 34 (1968): 44–50.

Glatthard, Thomas, and Alfred Bollinger. "Swiss Precision for U.S. Mapping: Ferdinand Rudolph Hassler, First Chief of U.S. Coast and Geodetic Survey and U.S. Bureau of Standards." *Proceedings of the FIG XXII International Congress,* Washington, D.C., April 19–26, 2002, section HS2, Surveying and Mapping the Americas—Great Surveyors. http://www.fig.net/pub/fig_2002/HS2/HS2_glatthard_bollinger.pdf.

Gore, Albert. *An Inconvenient Truth: The Planetary Emergency of Global Warming and What We Can Do About It.* Emmaus, Pa.: Rodale Press, 2006.

Graham, Douglas, Maryellen Sault, and Jonathan Bailey. "National Ocean Service Shoreline—Past, Present, and Future." *Journal of Coastal Research,* special issue 38 (Fall 2003): 14–32.

Greve, Ralf. "On the Response of the Greenland Ice Sheet to Greenhouse Climate Change." *Climatic Change* 46 (2000): 289–303.

Guthorn, Peter J. *United States Coastal Charts, 1783–1861.* Exton, Pa.: Schiffer Publishing, 1984.

Hansen, James E. "A Slippery Slope: How Much Global Warming Constitutes 'Dangerous Anthropogenic Interference'?" *Climatic Change* 68 (2005): 269–79.

Harrington, Charles E. "Maritime Boundaries on National Ocean Service Nautical Charts." *Cartographic Perspectives,* no. 14 (Winter 1993): 9–15.

Harris, D. L., N. A. Pore, and R. A. Cummings. "Tide and Tidal Current Prediction by High Speed Digital Computer." *International Hydrographic Review* 42 (January 1965): 95–103.

Harris, Elizabeth. "The Waldseemüller World Map: A Typographic Appraisal." *Imago Mundi* 37 (1985): 30–53.

Hassler, Ferdinand Rudolph. "Papers on Various Subjects Connected with the Survey of the Coast of the United States." *Transactions of the American Philosophical Society,* n.s., 2 (1825): 232–420.

Hearn, Chester G. *Tracks in the Sea: Matthew Fontaine Maury and the Mapping of the Oceans*. Camden, Maine: International Marine; New York: McGraw-Hill, 2002.

Heawood, Edward. "The Waldseemüller Facsimiles." *Geographical Journal* 23 (1904): 760–70.

Hebért, John R. "The Map That Named America: Library Acquires 1507 Waldseemüller Map of the World." *Library of Congress Information Bulletin* 62 (2003): 187–93.

Heffernan, Michael. "Geography, Cartography, and Military Intelligence: the Royal Geographical Society and the First World War." *Transactions of the Institute of British Geographers,* n.s. 21 (1996): 504–33.

———. "The Politics of the Map in the Early Twentieth Century." *Cartography and Geographic Information Science* 29 (2002): 207–26.

Henderson, Floyd M., and Anthony J. Lewis. "Introduction." In *Manual of Remote Sensing,* 3rd ed., edited by Floyd M. Henderson and Anthony J. Lewis, 2:1–7. New York: John Wiley and Sons, 1998.

Hess, Kurt W. "Tidal Datums and Tide Coordination." *Journal of Coastal Research,* special issue 38 (Fall 2003): 33–43.

Hessler, John W. "Warping Waldseemüller: A Phenomenological and Computational Study of the 1507 World Map." *Cartographica* 41, no. 2 (2006): 101–13.

Hicks, Steacy D. *The National Tidal Datum Convention of 1980*. Rockville, Md.: National Ocean Service, 1980.

Hinks, Arthur R. *Maps and Survey*. 4th ed. Cambridge: University Press, 1944.

———. "The Map on the Scale 1/1,000,000, Compiled at the Royal Geographical Society under the Direction of the General Staff, 1914–1915." *Geographical Journal* 46 (1915): 24–50.

———. "The 1/Million Map of Europe." *Geographical Journal* 94 (1939): 404–9.

Hiscock, Eric C. *Cruising under Sail*. 2nd ed. London: Oxford University Press, 1965.

Houston, Samuel H., and others. "Comparisons of HRD and SLOSH Surface Wind Fields in Hurricanes: Implications for Storm Surge Modeling." *Weather and Forecasting* 14 (1999): 671–86.

Intergovernmental Panel on Climate Change. *Climate Change 2007: The Physical Science Basis—Summary for Policymakers*. February 2007. Geneva, Switzerland.

International Civil Aviation Organization. *ICAO Aeronautical Chart Catalogue,* amendment no. 12. Montreal, July 30, 1965.

International Court of Justice. "Case Concerning Delimitation of the Maritime Boundary in the Gulf of Maine Area." *International Court of Justice Yearbook* 39 (1984–85): 246–390.

International Geographical Union. *Report of the Commission on the International Map of the World, 1:1,000,000*. New York: American Geographical Society, 1952.

International Map Committee. "International Map of the World." *Geographical Journal* 36 (1910): 179–84.

Jelesnianski, Chester P., Jye Chen, and Wilson A. Shaffer. *SLOSH: Sea, Lake, and*

Overland Surges from Hurricanes. NOAA Technical Report NWS 48. Silver Spring, Md.: National Weather Service, 1992.

Jenkins, Richard. "Social Construction." In *The Social Science Encyclopedia,* 3rd ed., edited by Adam Kuper and Jessica Kuper, 932–33. London: Routledge, 2004.

Jensen, John R. "Issues Involving the Creation of Digital Elevation Models and Terrain Corrected Orthoimagery Using Soft-Copy Photogrammetry." In *Digital Photogrammetry: An Addendum to the Manual of Photogrammetry,* edited by Cliff Greve, 167–79. Falls Church, Va.: American Society of Photogrammetry, 1996.

Joerg, W. L. G. "Development and State of Progress of the United States Portion of the International Map of the World." *Bulletin of the American Geographical Society* 44 (1912): 838–43.

Johnson, Douglas Wilson. "Is the Atlantic Coast Sinking?" *Geographical Review* 3 (1917): 135–39.

———. *Shore Processes and Shoreline Development.* New York: John Wiley and Sons, 1919.

Jones, Bennett G. "The Nine-Lens Air Camera in Use." *Bulletin of the Association of Field Engineers, U.S. Coast and Geodetic Survey,* no. 12 (December 1938): 120–25.

———. "Photogrammetric Surveys for Nautical Charts." *International Hydrographic Review* 32 (November 1955): 111–28.

Jones, E. Lester. *Elements of Chart Making.* U.S. Coast and Geodetic Survey Special Publication No. 38. Washington, D.C.: Government Printing Office, 1916.

———. "Evolution of the Nautical Chart." *Military Engineer* 16 (1924): 219–38.

Jones, Meirion T. "GEBCO Enters the Digital Era—the GEBCO Digital Atlas." In *The History of GEBCO, 1903–2003,* edited by Desmond Scott and others, 113–36. Lemmer, The Netherlands: Geomatics Information and Trading Center, 2003.

Karo, H. Arnold. "World Coastline Measurements." *International Hydrographic Review* 33 (May 1956): 131–40.

Kates, R. W., and others. "Reconstruction of New Orleans after Hurricane Katrina: A Research Perspective." *Proceedings of the National Academy of Sciences* 103 (2006): 14653–60.

Kerr, Adam J. "The International Hydrographic Bureau Period." In *The History of GEBCO, 1903–2003,* edited by Desmond Scott and others, 53–64. Lemmer, The Netherlands: Geomatics Information and Trading Center, 2003.

Kerr, Richard A. "A Worrying Trend of Less Ice, Higher Seas." *Science* 311 (2006): 1698–1700.

King, Rawle O. "Federal Flood Insurance: The Repetitive Loss Problem." Congressional Research Service Report RL32972, June 30, 2005. http://digital.library.unt.edu/govdocs/crs//data/2005/upl-meta-crs-7693/RL32972_2005Jun30.pdf.

Klein, Natalie S. *Dispute Settlement in the UN Convention on the Law of the Sea.* Cambridge: Cambridge University Press, 2005.

Kleinberg, Eliot. *Black Cloud: The Great Florida Hurricane of 1928*. New York: Carroll and Graf, 2003.

Knorr, E[rnest]. R[udolph]. *A Few Words on International Cooperation in Maritime Hydrography*. Washington, D.C.: Judd and Detweiler, 1884.

Knox, Robert W. "Mapping the Earth." *Journal of the Coast and Geodetic Survey*, no. 6 (April 1955): 65–74.

Koike, Nobuaki, and others. "Tsunami Run-up Heights of the 2004 Off the Kii Peninsula Earthquakes." *Earth, Planets and Space* 57 (March 2005): 157–60.

Konvitz, Josef. *Cartography in France, 1660–1848: Science, Engineering, and Statecraft*. Chicago: University of Chicago Press, 1987.

Kopec, Richard J. "Global Climate Change and the Impact of a Maximum Sea Level on Coastal Settlement." *Journal of Geography* 70 (1971): 541–50.

Kotch, Joseph A., Jr. "The 1:250,000-scale Map Series of the United States." M.A. paper, Syracuse University, 1982.

Larkin, Frank J. *Basic Coastal Navigation: An Introduction to Piloting*. Dobbs Ferry, N.Y.: Sheridan House, 1993.

Leatherman, Stephen P. "Shoreline Mapping: A Comparison of Techniques." *Shore and Beach* 51 (July 1983): 28–33.

———. "Social and Economic Costs of Sea Level Rise." In *Sea Level Rise: History and Consequences*, edited by Bruce C. Douglas, Michael S. Kearney, and Stephen P. Leatherman, 181–223. San Diego: Academic Press, 2001.

Leatherman, Stephen P., and Patricia Jones Kershaw. *Sea Level Rise and Coastal Disasters: Summary of a Forum, October 25, 2001*. Washington, D.C.: National Academy Press, 2002.

Leatherman, Stephen, and others. *Evaluation of Erosion Hazards: Summary*. Washington, D.C.: H. John Heinz III Center for Science, Economics and the Environment, 2000.

Lee County, Florida. Department of Community Development. *Lee County Coastal High Hazard Area* (map). http://www.lee-county.com/dcd1/downloads/maps/LeePlan/Map05.pdf (accessed February 22, 2006).

———. *The Lee Plan 2004 Codification, As Amended through December 2004*. http://lee-county.com/dcd1/leeplan/leeplan.pdf.

Lee County (Florida) Emergency Management. *Lee County Elevation and Evacuation Map*. http://www.leeeoc.com//evacroutes.cfm (accessed February 22, 2006).

Lee County, Florida. GIS Department. *Lee County Flood Zones* (map). http://www.leepa.org/GISDepartment/FIRM1.htm (accessed February 16, 2005).

Leighly, John. *California as an Island*. San Francisco: Book Club of California, 1972.

Lenczowski, Roberta A. "The Military as Users and Producers of Global Spatial Data." In *Framework for the World*, edited by David Rhind, 85–110. Cambridge: GeoInformation International, 1997.

Lethan, Lawrence. *GPS Made Easy: Using Global Positioning Systems in the Outdoors*. 4th ed. Seattle, Wash.: Mountaineers Books, 2003.

Library of Congress. *Universalis cosmographia secundum Ptholomaei traditionem et Americi Vespucii alioru[m]que lustrationes*. American Memory, http://memory.loc.gov.

Linklater, Andro. *Measuring America: How an Untamed Wilderness Shaped the United States and Fulfilled the Promise of Democracy*. New York: Walker and Co., 2002.

Liu, Hongxing, and Kenneth C. Jazek. "A Complete High-Resolution Coastline of Antarctica Extracted from Orthorectified Radarsat SAR Imagery." *Photogrammetric Engineering and Remote Sensing* 70 (2004): 605–16.

Loeper, Thomas J., and Jason J. Shadid. "Digital Update Service and Print-on-Demand." Paper presented at the 1999 U.S. Hydrographic Conference, Mobile, Alabama, April 27–29, 1999. Online at http://www.thsoa.org/us99papers.htm.

Loomer, Scott A. "Mathematical Analysis of Medieval Sea Charts." *Technical Papers, 1986 ACSM-ASPRS Annual Convention*, 1:123–32.

Lowe, Anthony S. "The Federal Emergency Management Agency's Multi-Hazard Flood Map Modernization and *The National Map*." *Photogrammetric Engineering and Remote Sensing* 69 (2003): 1133–35.

Macey, Samuel L. *The Dynamics of Progress: Time, Method, and Measure*. Athens: University of Georgia Press, 1989.

Macleod, M. N. "The International Map." *Geographical Journal* 66 (1925): 445–49.

Maling, D. H. "How Long Is a Piece of String?" *Cartographic Journal* 5 (1968): 147–56.

———. *Measurements from Maps: Principles and Methods of Cartometry*. Oxford: Pergamon Press, 1989.

Manning, Thomas G. *U.S. Coast Survey vs. Naval Hydrographic Office: A 19th-Century Rivalry in Science and Politics*. Tuscaloosa: University of Alabama Press, 1988.

Marmer, H. A. "Sea Level along the Atlantic Coast of the United States and Its Fluctuations." *Geographical Review* 15 (1925): 438–48.

Massey, William G., and others. "History of Inundation Models." *Marine Technology Society Journal* 41 (Spring 2007): 7–17.

Martin, Geoffrey J. *The Life and Thought of Isaiah Bowman*. Hamden: Conn.: Archon Books, 1980.

Maury, Matthew Fontaine. "Blank Charts on Board Public Cruisers." *Southern Literary Messenger* 9 (1943): 458–61.

———. "Steam Navigation to China." *Southern Literary Messenger* 14 (1948): 246–54.

Mays, James. "Thorough and Versatile Tide Prediction by Computer." *Sea Technology* 28 (May 1987): 33–37.

McCurdy, P. G. *Manual of Coastal Delineation from Aerial Photographs*. H.O. Publication no. 592. Washington: Hydrographic Office, U.S. Navy, 1947.

McLaughlin, Glen, and Nancy H. Mayo. *The Mapping of California as an Island: An Illustrated Checklist*. Saratoga, Calif.: California Map Society, 1995.

Mercado, Aurelio. "On the Use of NOAA's Storm Surge Model, SLOSH, in Managing Coastal Hazards—The Experience in Puerto Rico." *Natural Hazards* 10 (1994): 235–46.

Merriman, Mansfield. *Elements of Precise Surveying and Geodesy*. 2nd ed., revised. New York: John Wiley and Sons, 1908.

Mills, Jerry, and Stephen Gill. "Water Levels and Flow." In *Manual of Hydrography,* International Hydrographic Organization, 253–99. Monaco: International Hydrographic Bureau, 2005.

Minerals Management Service. *Implementation Plan for a Multipurpose Marine Cadastre*. Washington, D.C., July 2004.

Monkhouse, F. J., and H. R. Wilkinson. *Maps and Diagrams: Their Compilation and Construction*. London: Methuen and Co., 1952.

Monmonier, Mark. "Practical and Emblematic Roles of the American Polyconic Projection." *Weiner Schriften zur Geographie und Kartographie* [Institut für Geographie und Regionalforschung der Universität Wien] 16 (2004): 93–99.

———. *Rhumb Lines and Map Wars: A Social History of the Mercator Projection*. Chicago: University of Chicago Press, 2004.

Moore, Laura J. "Shoreline Mapping Techniques." *Journal of Coastal Research* 16 (2000): 111–24.

Morison, Samuel Eliot. *Admiral of the Ocean Sea: A Life of Christopher Columbus*. 2 vols. Boston: Little, Brown and Co., 1942.

———. "Route of Columbus along the North Coast of Haiti and the Site of Navidad." *Transactions of the American Philosophical Society* 30 (1940): 239–85.

———. "Texts and Translations of the Journal of Columbus's First Voyage." *Hispanic American Historical Review* 19 (1939): 235–61.

Mroz, Monica, and others. "The Defense Mapping Agency (DMA) World Vector Shoreline Plus (WVSPLUS) Revision and Conversion Project." *Proceedings of the 1996 ESRI International User Conference, Palm Springs, California, May 20–24, 1996,* http://gis.esri.com/library/userconf/proc96/TO250/PAP237/P237.HTM.

Nassar, Haya el-. "Storms Don't Deter N.C. Coastal Residents." *USA Today,* October 21, 2005.

National Geodetic Survey. "Importance of Shoreline" (diagram). http://www.ngs.noaa.gov/RSD/coastal/importance.html.

National Geospatial-Intelligence Agency. "NGA Raster Roam." http://geoengine.nga.mil/geospatial/SW_TOOLS/NIMAMUSE/webinter/rast_roam.html.

National Geospatial-Intelligence Agency. Maritime Safety Division. "Using Nautical Charts with Global Positioning System, Edition 3." Updated March 2, 2004. http://pollux.nss.nima.mil/index/uncon/dnc_gps/.

National Oceanic and Atmospheric Administration. *The Coastline of the United States*. NOAA/PA 71046 (Rev 1975). Washington, D.C.: U.S. Department of Commerce, 1975.

National Oceanic and Atmospheric Administration. Center for Operational Oceanic Products and Services. *Computational Techniques for Tidal Datums Handbook*. NOAA Special Publication NOS CO-OPS 2. Silver Spring, Md.: National Ocean Service, 2003.

———. "Predicted Water Level Data." http://co-ops.nos.noaa.gov.

———. "Sea Levels Online." http://tidesandcurrents.noaa.gov/sltrends/sltrends.shtml.

———. "Tide-Predicting Machine No. 2." http://co-ops.nos.noaa.gov/predma2.html.

National Oceanic and Atmospheric Administration. Coastal Services Center. "World Vector Shoreline Plus." Updated August 1, 2005. http://www.csc.noaa.gov/shoreline/nga_wvsplus.html.

National Oceanic and Atmospheric Administration. National Geodetic Survey. Remote Sensing Division. "CSCAP: Coast and Shoreline Change Analysis Program." Modified December 19, 2003. http://www.ngs.noaa.gov/RSD/coastal/cscap.shtml.

———. "Imaging Spectroscopy (Hyperspectral)." Modified May 21, 2003. http://www.ngs.noaa.gov/RESEARCH/RSD/main/hyper/hyper.shtml.

———. "Synthetic Aperture Radar (SAR)." Modified May 20, 2003. http://www.ngs.noaa.gov/RESEARCH/RSD/main/sar/sar.shtml.

National Oceanic and Atmospheric Administration. National Hurricane Center. "The Saffir-Simpson Hurricane Scale." Modified August 22, 2005. http://www.nhc.noaa.gov/aboutsshs.shtml.

National Oceanic and Atmospheric Administration. National Ocean Service. *Coast Pilot 1—Atlantic Coast: Eastport to Cape Cod.* 35th ed. Washington, D.C.: U.S. Department of Commerce, 2005.

National Oceanic and Atmospheric Administration. National Weather Service. Evaluation Branch. "Hurricane Storm Surge Forecasting." http://www.weather.gov/mdl/marine/Basin.html (accessed February 1, 2007).

National Oceanic and Atmospheric Administration. Office of Coast Survey. "Chart Accuracy." http://chartmaker.ncd.noaa.gov/staff/Accuracy.htm.

———. *Chart No. 1: Nautical Chart Symbols, Abbreviations, and Terms.* 10th ed. November 1997. http://chartmaker.ncd.noaa.gov/mcd/chart1/chart1hr.htm.

———. "Frequently Asked Questions." http://chartmaker.ncd.noaa.gov/staff/faq.htm#sources.

———. "Historical Maps and Charts—Background." Revised April 9, 2002. http://nauticalcharts.noaa.gov/csdl/ctp/backg.htm.

———. "Navigational Charts." Revised August 31, 2005. http://nauticalcharts.noaa.gov/staff/charts.htm.

National Research Council. Committee on Abrupt Climate Change. *Abrupt Climate Change: Inevitable Surprises.* Washington, D.C.: National Academy Press, 2002.

National Research Council. Committee on Coastal Erosion Zone Management. *Managing Coastal Erosion.* Washington, D.C.: National Academy Press, 1990.

National Research Council. Committee on Coastal Flooding from Hurricanes. *Evaluation of the FEMA Model for Estimating Potential Coastal Flooding from Hurricanes and Its Application to Lee County, Florida.* Washington, D.C.: National Academy Press, 1983.

National Research Council. Committee on Engineering Implications of Changes in Relative Mean Sea Level. *Responding to Changes in Sea Level: Engineering Implications.* Washington, D.C.: National Academy Press, 1987.

National Research Council. Committee on Floodplain Mapping Techniques.

Base Map Inputs for Floodplain Mapping. Washington, D.C.: National Academies Press, 2007.

National Research Council. Committee on National Needs for Coastal Mapping and Charting. *A Geospatial Framework for the Coastal Zone: National Needs for Coastal Mapping and Charting*. Washington, D.C.: National Academies Press, 2004.

National Research Council. Committee on Options for Preserving Cape Hatteras Lighthouse. *Saving Cape Hatteras Lighthouse from the Sea: Options and Policy Implications*. Washington, D.C.: National Academy Press, 1988.

National Research Council. Committee on Planning for Catastrophe: A Blueprint for Improving Geospatial Data, Tools, and Infrastructure. *Successful Response Starts with a Map: Improving Geospatial Support for Disaster Management*. Washington, D.C.: National Academies Press, 2007.

Nebenzahl, Kenneth. *Atlas of Columbus and the Great Discoveries*. Chicago: Rand McNally, 1990.

Nelson, Harold E., and Joseph F. Dracup. "A Tale of Two Eras in American Geodesy." *Surveying and Land Information Systems* 60 (2000): 163–76.

Nichols, C. L. "International Co-operation and Co-ordination in Cartography as Viewed by Hydrographers." *World Cartography* 3 (1953): 13–17.

Nordenskiöld, Adolf Erik. *Facsimile Atlas to the Early History of Cartography with Reproductions of the Most Important Maps Printed in the XV and XVI Centuries*, trans. Johan Adolf Ekelöf and Clements R. Markham. Stockholm, 1889; New York: Dover Publications, 1973.

Nowatzki, John, and others. "GPS Applications in Crop Production." North Dakota State University Extension Service Bulletin AE-1264. April 2004. http://www.ext.nodak.edu/extpubs/ageng/gis/ae1264w.htm.

Odgers, Merle M. *Alexander Dallas Bache, Scientist and Educator, 1806–1867*. Philadelphia: University of Pennsylvania Press, 1947.

Ogden, Herbert G. "The Survey of the Coast." *National Geographic Magazine* 1 (1888): 59–77.

Pajak, Mary Jean, and Stephen Leatherman. "The High Water Line as Shoreline Indicator." *Journal of Coastal Research* 18 (2002): 329–37.

Parizek, Byron R., and Richard B. Alley. "Implications of Increased Greenland Surface Melt under Global-Warming Scenarios: Ice-Sheet Simulations." *Quaternary Science Reviews* 23 (2004): 1013–27.

Parkhurst, Douglas L. "Salient Features of the Design of a New Tide-Predicting Machine." *International Hydrographic Review* 37 (July 1960): 105–11.

Patton, Raymond Stanton. "Recent Advancements in Coast and Geodetic Survey Methods." *Annals of the Association of American Geographers* 22 (1932): 1–11.

Pearcy, G. Etzel. "Hawaii's Territorial Sea." *Professional Geographer* 11 (November 1959): 2–6.

———. "Measurement of the U.S. Territorial Sea." *Department of State Bulletin* 40 (1959): 963–71.

Penck, A[lbrecht]. "The Construction of a Map of the World on a Scale of 1:1,000,000." *Geographical Journal* 1 (1893): 253–61.

Perugini, Nick. "Behind the Accuracy of Electronic Charts." *Sea Technology* 42 (March 2001): 33–37.

Pilkey, Orrin H., and others. *The North Carolina Shore and Its Barrier Islands: Restless Ribbons of Sand*. Durham, N.C.: Duke University Press, 1998.

Platt, Raye R. "The Map of Hispanic America on the Scale of 1:1,000,000." *Geographical Review* 36 (1946): 1–28.

Plumlee, Matthew D., and others. "The CCOM Chart-of-the-Future Project: Maximizing Mariner Effectiveness through Fusion of Marine and Visualization Technologies." Paper presented at the 7th Marine Transportation System Research and Technology Coordination Conference, Washington, D.C., November 16–17, 2004. Online at http://www.trb.org/Conferences/MTS/2A%20PlumleePaper.pdf.

Polk, Dora Beale. *The Island of California: A History of the Myth*. Spokane, Wash.: Arthur H. Clark, 1991.

Posner, Richard A. *Catastrophe: Risk and Response*. New York: Oxford University Press, 2004.

Prescott, Victor, and Clive Schofield. *The Maritime Political Boundaries of the World*. 2nd ed. Leiden: Martinus Nijhoff, 2005.

Psuty, Norbert P., and Douglas D. Ofiara. *Coastal Hazard Management: Lessons and Future Directions from New Jersey*. New Brunswick, N.J.: Rutgers University Press, 2002.

Putnam, G. R. *Nautical Charts*. New York: John Wiley and Sons, 1908.

Randall, Martin M. "Coastal Development Run Amuck: A Policy of Retreat May Be the Only Hope." *Journal of Environmental Law and Litigation* 18 (2003): 145–85.

Reading, O. S. "Aerial Photography and Coast Surveys." *Bulletin of the Association of Field Engineers, U.S. Coast and Geodetic Survey*, no. 3 (June 1931): 9–12.

———. "The Nine Lens Aerial Camera of the Coast and Geodetic Survey." *Bulletin of the Association of Field Engineers, U.S. Coast and Geodetic Survey*, no. 9 (December 1935): 56–62.

Reagan, Ronald (President of the United States). "Proclamation 5030 of March 10, 1983: Exclusive Economic Zone of the United States of America." *Federal Register* 48 (March 14, 1983): 10605–6.

———. "Proclamation 5928 of December 27, 1988: Territorial Sea of the United States of America." *Federal Register* 54 (January 9, 1989): 777.

Reed, Michael W. *The Development of International Maritime Boundary Principles through United States Practice*. Vol. 3 of *Shore and Sea Boundaries*. Washington, D.C.: Government Printing Office, 2000.

Rennell, Lord, Isaiah Bowman, and Lord Halifax. "The Geographical Situation of the United States in Relation to World Policies: Discussion." *Geographical Journal* 112 (1948): 142–45.

Rhind, David. "Current Shortcomings of Global Mapping and the Creation of a New Geographical Framework for the World." *Geographical Journal* 166 (2000): 295–305.

Riddle, Lyn. "Crawling, a Lighthouse Flees the Battering Sea." *New York Times*, June 18, 1999.

Ristow, Walter W. *Aviation Cartography: A Historico-Bibliographic Study of Aeronautical Charts*. Washington, D.C.: Library of Congress, 1960.

Ritchie, George Stephen. *Challenger: The Life of a Survey Ship*. New York and London: Abelard-Schuman, 1958.

———. Introduction to Derek Howse and Michael Sanderson's *The Sea Chart: An Historical Survey Based on the Collections in the National Maritime Museum*, 9–13. New York: McGraw-Hill, 1973.

Rizos, Chris. "Introducing the Global Positioning System." In *Manual of Geospatial Science and Technology*, edited by John D. Bossler, 77–94. New York: Taylor and Francis, 2002.

Robertson, William, V, Dean Whitman, Keqi Zhang, and Stephen P. Leatherman. "Mapping Shoreline Position Using Airborne Laser Altimetry." *Journal of Coastal Research* 20 (2004): 884–92.

Robinson, A. H. W. *Marine Cartography in Britain: A Story of the Sea Chart to 1855*. Leicester, U.K.: Leicester University Press.

Robinson, Arthur H. "The Future of the International Map." *Cartographic Journal* 2 (1967): 23–26.

Schilder, Günter. "Willem Jansz. Blaeu's Wall Map of the World, on Mercator's Projection, 1606–7 and Its Influence." *Imago Mundi* 31 (1979): 36–54.

Schneider, Stephen H., and Robert S. Chen. "Carbon Dioxide Warming and Coastline Flooding: Physical Factors and Climatic Impact." *Annual Review of Energy* 5 (1980): 107–40.

Schott, C. A. "Recent Contributions to Our Knowledge of the Earth's Shape and Size, by the United States Coast and Geodetic Survey." *National Geographic Magazine* 12 (1901): 36–41.

Schureman, Paul. *Manual of Harmonic Analysis and Prediction of Tides*. U.S. Coast and Geodetic Survey Special Publication No. 98, revised. Washington, D.C.: Government Printing Office, 1958.

Schwartz, Charles R. "North American Datum of 1927." In *North American Datum of 1983*, edited by Charles R. Schwartz, 5–8. NOS Professional Paper NOS 2. Rockville, Md.: National Geodetic Survey, 1989.

Scott, Desmond P. D. "A Change of Direction." In *The History of GEBCO, 1903–2003*, edited by Desmond Scott and others, 95–106. Lemmer, The Netherlands: Geomatics Information and Trading Center, 2003.

Scott, Desmond P. D., and others, eds. *The History of GEBCO 1903–2003*. Lemmer, The Netherlands: Geomatics Information and Trading Center, 2003.

Shalowitz, Aaron L. *Interpretation and Use of Coast and Geodetic Survey Data*. Vol. 2 of *Shore and Sea Boundaries*, Coast and Geodetic Survey Publication 10-1. Washington, D.C.: Government Printing Office, 1964.

———. "Safeguarding Our Seaways—the Modern Nautical Chart." *Scientific Monthly* 61 (1945): 249–64.

Shelley, Anthony F. "Law of the Sea: Delimitation of the Gulf of Maine." *Harvard International Law Journal* 26 (1985): 646–54.

Shen, Jian, and others. "Improved Prediction of Storm Surge Inundation with a High-Resolution Unstructured Grid Model." *Journal of Coastal Research* 22 (2006): 1309–19.

Shirley, Rodney W. *The Mapping of the World: Early Printed World Maps, 1472–1700.* London: Holland Press, 1983.

Silliman, Benjamin. "The Coast Survey of the United States." *American Journal of Science* 49 (1845): 229–49.

Simmons, Lansing G. "Geometric Techniques in Geodesy." In *Contemporary Geodesy: Proceedings of a Conference Held at the Harvard College Observatory—Smithsonian Astrophysical Observatory, Cambridge, Massachusetts, December 1–2, 1958*, edited by Charles A. Whitten and Kenneth H. Drummond, 4–6. Geophysical Monograph No. 4. Washington, D.C.: American Geophysical Union, 1959.

Skelton, R. A. "The Cartography of Columbus' First Voyage." In *The Journal of Christopher Columbus*, translated by Cecil Jane, 217–27. New York: Clarkson N. Potter, 1960.

Slotten, Hugh Richard. *Patronage, Practice, and the Culture of American Science: Alexander Dallas Bache and the U.S. Coast Survey.* Cambridge: Cambridge University Press, 1994.

Small, Christopher, and Robert J. Nicholls. "A Global Analysis of Human Settlement in Coastal Zones." *Journal of Coastal Research* 19 (2003): 584–99.

Smith, James R. *Introduction to Geodesy: The History and Concepts of Modern Geodesy.* New York: John Wiley and Sons, 1997.

———. "The Pear-Shaped Earth." *Geographical Magazine* 58 (1986): 572–77.

Smith, John T., Jr. *A History of Flying and Photography in the Photogrammetry Division of the National Ocean Survey, 1919–79.* Rockville, Md.: National Ocean Service, 1981.

Smith, Neil. *American Empire: Roosevelt's Geographer and the Prelude to Globalization.* Berkeley and Los Angeles: University of California Press, 2003.

Smith, Robert W. "The Maritime Boundaries of the United States." *Geographical Review* 71 (1981): 395–410.

Snyder, John P. *Flattening the Earth: Two Thousand Years of Map Projections.* Chicago: University of Chicago Press, 1993.

———. "The Modified Polyconic Projection for the IMW." *Cartographica* 19 (Autumn and Winter 1982): 31–43.

Sobel, Dava. *Longitude: The True Story of a Lone Genius Who Solved the Greatest Scientific Problem of His Time.* New York: Walker and Company, 1995.

Soluri, E. A., and V. A. Woodson. "World Vector Shoreline." *International Hydrographic Review* 67 (January 1990): 27–35.

Spilhaus, Athelstan F. "Maps of the Whole World Ocean." *Geographical Review* 32 (1942): 431–35.

Stachurski, Richard J. "Longitude by Wire: The American Method." *Professional Surveyor Magazine* 23 (November 2003): 16–24.

Steinberg, Philip E. *The Social Construction of the Ocean.* Cambridge: Cambridge University Press, 2001.

Stem, James E. "User Participation and Impact." In *North American Datum of 1983*, edited by Charles R. Schwartz, 237–48. NOS Professional Paper NOS 2. Rockville, Md.: National Geodetic Survey, 1989.

Stembel, Oren E. "Horizontal Datum Anomalies on Nautical Charts: A Solution." In *Proceedings of the International Symposium on Marine Positioning, Res-*

ton, VA, October 14–17, 1986*, edited by Muneendra Kumar and George A. Maul, 265–73. Dordrecht, Holland: D. Reidel, 1987.

Stembel, Oren E., Jr., and William J. Monteith. *Implementation of North American Datum of 1983 into the NOS Nautical Charting Program*. NOAA Report NOS 115, CGS 8. Rockville, Md.: National Oceanic and Atmospheric Administration, 1985.

Stevenson, Edward Luther. "Martin Waldseemüller and the Early Lusitano-Germanic Cartography of the New World." *Bulletin of the American Geographical Society* 36 (1904): 192–215.

———. *Portolan Charts: Their Origin and Characteristics with a Descriptive List of Those Belonging to the Hispanic Society of America*. New York: Knickerbocker Press, 1911.

Stockdon, Hilary F., et al. "Estimation of Shoreline Position and Change Using Airborne Topographic Lidar Data." *Journal of Coastal Research* 18 (2002): 502–13.

Stommel, Henry. *Lost Islands: The Story of Islands That Have Vanished from Nautical Charts*. Vancouver: University of British Columbia Press, 1984.

Stone, Gregory W., Xiongping Zhang, and Alexandru Sheremet. "The Role of Barrier Islands, Muddy Shelf and Reefs in Mitigating the Wave Field along Coastal Louisiana." *Journal of Coastal Research*, special issue 44 (Spring 2005): 40–55.

Suárez, Thomas. *Early Mapping of the Pacific*. Singapore: Periplus Editions, 2004.

Swainson, O. W. *Topographic Manual*. U.S. Coast and Geodetic Survey Special Publication No. 144. Washington, D.C.: Government Printing Office, 1928.

Taylor, E. G. R., and M. W. Richey. *The Geometrical Seaman: A Book of Early Nautical Instruments*. London: Hollis and Carter, 1962.

Tewinkel, G. C. "Stereoplotter for Nine-Lens Photographs." *Journal of the Coast and Geodetic Survey*, no. 1 (January 1948): 45–50.

Thatcher, John Boyd. *Christopher Columbus: His Life, His Works, His Remains*. 3 vols. New York: G. P. Putnam's Sons, 1903–4.

Theberge, Albert E. *The Coast Survey: 1807–1867*. Vol. 1, *The History of the Commissioned Corps of the National Oceanic and Atmospheric Administration*. Silver Spring, Md.: National Oceanic and Atmospheric Administration, 1998. Available only online at http://www.lib.noaa.gov/edocs/heritage.html.

———. "The United States Coast Pilot—A Short History." Office of Coast Survey, http://nauticalcharts.noaa.gov/nsd/cphistory.htm.

Thompson, Morris M. *Maps for America: Cartographic Products of the U.S. Geological Survey and Others*. Washington, D.C.: Government Printing Office, 1979.

Thompson, Morris M., and Heinz Gruner. "Foundations of Photogrammetry." In *Manual of Photogrammetry*, 4th ed., edited by Chester C. Slama, Charles Theurer, and Soren W. Henriksen, 1–36. Falls Church, Va.: American Society of Photogrammetry, 1980.

"The Times" Atlas. London: Office of "The Times," 1896.

Titus, James G. "Does the U.S. Government Realize That the Sea Is Rising? How to Restructure Federal Programs So That Wetlands and Beaches Survive." *Golden Gate University Law Review* 30 (2000): 717–78.

———. "Maps That Depict the Business-As-Usual Response to Sea Level Rise in the Decentralized United States of America." OECD (Organisation for Economic Co-operation and Development) Global Forum on Sustainable Development: Development and Climate Change, Paris. November 11–12, 2004.

———. "Rising Seas, Coastal Erosion, and the Takings Clause: How to Save Wetlands and Beaches without Hurting Property Owners." *Maryland Law Review* 57 (1998): 1279–1399.

Titus, James G., and Vijay Narayanan. "The Risk of Sea Level Rise." *Climatic Change* 33 (1996): 151–212.

Titus, James G., and Charlie Richman. "Maps of Lands Vulnerable to Sea Level Rise: Modeled Elevations along the U.S. Atlantic and Gulf Coasts." *Climate Research* 18 (2001): 205–28.

Titus, James G., and others. "Greenhouse Effect and Sea Level Rise: The Cost of Holding Back the Sea." *Coastal Management* 19 (1991): 171–204.

Tooley, R. V. *California as an Island: A Geographical Misconception Illustrated by 100 Examples from 1625 to 1770*. Map Collectors' Series No. 8. London: Map Collectors' Circle, 1964.

Townsend, Francis Fragos. *The Federal Response to Hurricane Katrina: Lessons Learned*. Washington, D.C.: The White House, 2006. http://www.whitehouse.gov/reports/katrina-lessons-learned/.

Tucker, Compton J., Denelle M. Grant, and Jon D. Dykstra. "NASA's Global Orthorectified Landsat Data Set." *Photogrammetric Engineering and Remote Sensing* 70 (2004): 313–22.

Turner, John F. "Accession to the 1982 Law of the Sea Convention and Ratification of the 1994 Agreement Amending Part XI of the Law of the Sea Convention." Testimony before the Senate Environment and Public Works Committee, March 23, 2004. Senate Treaty Document 103-39 and Senate Executive Report 108-10. Online at http://www.state.gov/g/oes/rls/rm/2004/30723.htm.

United Nations. Secretariat. International Map of the World on the Millionth Scale: Report for 1955. New York: United Nations Department of Economic and Social Affairs, 1957.

———. "The International Map of the World on the Millionth Scale and the International Co-operation in the Field of Cartography." *World Cartography* 3 (1953): 1–13.

———. "Status of Hydrographic Surveying and Nautical Charting Worldwide." *World Cartography* 22 (1993): 1–35.

United Nations. Division for Ocean Affairs and the Law of the Sea. Chronological Lists of Ratifications, Accessions and Successions to the Convention and the Related Agreements. http://www.un.org/Depts/los/reference_files/chronological_lists_of_ratifications.htm.

———. Oceans and Laws of the Sea. http://www.un.org/Depts/los/convention_agreements/convention_overview_convention.htm.

United Nations. Office for Ocean Affairs and the Law of the Sea. *The Law of the Sea—Baselines: National Legislation with Illustrative Maps*. New York, 1989.

U.S. Army Corps of Engineers. Interagency Performance Evaluation Task Force. *Performance Evaluation of the New Orleans and Southeast Louisiana Hurricane Protection System, Draft Final Report, 1 June 2006*. Volume I—Executive Summary and Overview. https://ipet.wes.army.mil/ (accessed August 15, 2006).

U.S. Army Corps of Engineers. Philadelphia District. "Hazards and Vulnerability Data." Chapter 2 in *Hurricane Opal Assessment: Review of the Use and Value of Hurricane Evacuation Studies in the Hurricane Opal Evacuation, Alabama and Florida, October 3–4, 1995*. U.S. Army Corps of Engineers, Mobile District, and Federal Emergency Management Agency, Region IV, September 1966. http://www.csc.noaa.gov/hes/docs/postStorm/H_OPAL_ASSESMENT_REVIEW_USE_VALUE_HES_PRODUCTS_OPAL_EVACUATION_AL_FL.pdf.

U.S. Coast and Geodetic Survey. *Annual Report of the Director, United States Coast and Geodetic Survey to the Secretary of Commerce for the Fiscal Year Ended June 30, 1930*. Washington, D.C.: Government Printing Office, 1930.

———. *Atlantic Local Coast Pilot, Sub-Division 4: White Head Island to Cape Small Point*. Washington, D.C.: Government Printing Office, 1879.

———. *Coastline of the United States*. July 1948.

———. *Description of the U.S. Coast and Geodetic Survey Tide-Predicting Machine No. 2*. Special Publication No. 32. Washington: Government Printing Office, 1915.

———. *General Instructions for the Field Work of the U.S. Coast and Geodetic Survey*. Special Publication No. 26. Washington: Government Printing Office, 1915.

———. *Lengths, in Statute Miles, of the General Coast Line and Tidal Shore Line of the United States and Outlying Territories*. Serial 22. Washington: Government Printing Office, 1915.

———. *The United States Coast and Geodetic Survey. Its Work, Methods, and Organization*. Special Publication No. 23. Washington, D.C.: Government Printing Office, 1828.

U.S. Coast Guard Navigation Center. "DGPS General Information." Updated July 26, 2005. http://www.navcen.uscg.gov/dgps/.

U.S. Coast Survey. *Coast-Pilot for the Atlantic Sea-Board, Gulf of Maine and Its Coast from Eastport to Boston, 1874*. Washington: Government Printing Office, 1875.

U.S. Congress. House Committee on Foreign Relations. *Three Treaties Establishing Maritime Boundaries between the United States and Mexico, Venezuela, and Cuba*. 96th Cong., 2d sess., 1980. Senate Executive Rept. 96-49.

U.S. Congress. Senate Committee on Foreign Relations. *U.S.–Mexico Treaty on Maritime Boundaries*. 105th Cong., 1st sess., 1997. Senate Executive Rept. 105-4.

U.S. Congress. Senate Committee on Naval Affairs. *Steam Communication with China and the Sandwich Islands*. 30th Cong., 1st sess., 1848. Rept. 596.

U.S. Defense Mapping Agency. *Geodesy for the Layman*. Technical Report 80-003. Washington, D.C.: Defense Mapping Agency, 1983. http://handle.dtic.mil/100.2/ADA142764.

U.S. Department of Defense. Undersecretary of Defense for Policy. *Maritime Claims Reference Manual*, DoD 2005.1-M, June 23, 2005. http://www.dtic.mil/whs/directives/corres/html/20051m.htm.

U.S. Department of the Navy. Chief Information Officer. *DONCIO Term Glossary*, http://www.don-imit.navy.mil/glossary/default.asp (accessed December 27, 2005).

U.S. Department of State. *International Map of the World: Communication from the President of the United States Transmitting a Report from the Secretary of State Recommending a Request to Congress for an Appropriation for a Contribution by the United States toward the Expenses of the Bureau for the International Map of the World*. 68th Cong., 2d sess., December 30, 1924, Senate Document 177.

U.S. Department of State. Bureau of Oceans and International Environmental and Scientific Affairs. "Maritime Boundary: Cuba–United States." *Limits in the Seas*, no. 110. February 21, 1990. http://www.state.gov/documents/organization/58380.pdf.

———. "United States Responses to Excessive National Maritime Claims." *Limits in the Seas*, no. 112. March 9, 1992. http://www.state.gov/documents/organization/58381.pdf.

U.S. Department of State. Bureau of Public Affairs. Treaties in Force 2005. http://www.state.gov/documents/organization/53724.pdf (accessed July 19, 2006).

U.S. Department of State. Office of the Geographer. *Sovereignty of the Sea*. Geographic Bulletin 3. Washington, D.C.: Government Printing Office, 1965.

U.S. Environmental Protection Agency. Office of Air and Radiation. Forthcoming Products. January 2002. http://yosemite.epa.gov/oar/globalwarming.nsf/content/ResourceCenterPublicationsSLRProducts.html (accessed April 27, 2006).

———. Maps of Lands Vulnerable to Sea Level Rise. July 2000. http://yosemite.epa.gov/OAR/globalwarming.nsf/content/ResourceCenterPublicationsSLRMaps.html (accessed April 27, 2006).

U.S. General Accounting Office. *National Flood Insurance: Marginal Impact on Flood Plain Development, Administrative Improvements Needed*. GAO Report CED-82-105, August 16, 1982. http://archive.gao.gov/d44t15/119428.pdf.

———. *National Flood Insurance Program: Actions to Address Repetitive Loss Properties*. (Testimony of William O. Jenkins, Jr., Director, Homeland Security & Justice Issues, GAO.) Report GAO-04-401T, March 25, 2004. http://www.gao.gov/new.items/d04401t.pdf. (Accessed February 15, 2006).

U.S. Geological Survey. "USEEZ: Boundaries of the Exclusive Economic Zones of the U.S. and Territories." May 12, 2005. http://coastalmap.marine.usgs.gov/GISdata/basemaps/boundaries/eez/NOAA/useez_noaa.htm.

U.S. Government Accountability Office. *U.S Tsunami Preparedness: Federal and State Partners Collaborate to Help Communities Reduce Potential Impacts, but Significant Challenges Remain*. Report GAO-06-519, June 2006.

U.S. Supreme Court. *United States v. Alaska*. 521 U.S. 1 (1997).

Verner, Coolie. "Copperplate Engraving." In *Five Hundred Years of Map Printing*, edited by David Woodward, 51–75. Chicago: University of Chicago Press, 1975.

Vidal, Susan, Doug Graham, and Maryellen Sault. "CSCAP: Coast and Shoreline Change Analysis Program; Using High-Resolution Satellite Imagery for Shoreline Change Evaluation within Ports." U.S. Hydrographic Conference 2001, Norfolk, Va. May 22–24, 2001. http://www.thsoa.org/hy01/5_2.pdf.

Vielé, Egbert Ludovickus. *The Topography and Hydrology of New York*. New York: R. Craighead, 1865.

Wade, Elizabeth B. "Impact of North American Datum of 1983." *Journal of Survey Engineering* 112 (1986): 49–62.

Wainwright, D. B. *Plane Table Manual*. U.S. Coast and Geodetic Survey Special Publication No. 85. Washington, D.C.: Government Printing Office, 1922.

Warita, Yoshio, and Kunio Nonomura. "The National and Global Activities of the Japanese National Mapping Organization." In *Framework for the World*, edited by David Rhind, 31–47. Cambridge: GeoInformation International, 1997.

Warner, Deborah Jean. "Political Geodesy: The Army, the Air Force, and the World Geodetic System of 1960." *Annals of Science* 59 (2002): 363–89.

Weber, Gustavus A. *The Coast and Geodetic Survey: Its History, Activities and Organization*. Baltimore: Johns Hopkins Press, 1923.

———. *The Hydrographic Office: Its History, Activities and Organization*. Baltimore: Johns Hopkins Press, 1926.

Weddle, Kevin J. "The Blockade Board of 1861 and Union Naval Strategy." *Civil War History* 48 (2002): 123–42.

Wells, David. "Representing Uncertainty on Charts: The Hydrographic Crisis." *Sea Technology* 45 (June 2004): 89.

Wharton, William J. L. *Hydrographic Surveying: A Description of the Means and Methods Employed in Constructing Marine Charts*. 2nd ed., rev. London: John Murray, 1898.

Wheeler, W. H. *A Practical Manual of Tides and Waves*. New York: Longmans, Green, and Co., 1906.

Whitfield, Peter. *The Charting of the Oceans: Ten Centuries of Maritime Maps*. Rhonert Park, Calif.: Pomegranate Artbooks, 1996.

Whiting, Henry L. "Reports of Assistant H. L. Whiting on Topographical Contour, Hydrographic Details, and Reduction, on Photography, and on the Scale of Shades Suitable for Complete Maps." Appendix 20 in *Report of the Superintendent of the Coast Survey Showing the Progress of the Survey during the Year 1860*, 216–29. Washington, D.C.: Government Printing Office, 1861.

Williams, Francis Leigh. *Matthew Fontaine Maury: Scientist of the Sea*. New Brunswick, N.J.: Rutgers University Press, 1963.

Willis, Bailey. "The International Millionth Map of the World." *National Geographic Magazine* 21 (1910): 125–32.

Wilson, Herbert M. *Topographic, Trigonometric and Geodetic Surveying*. 3rd ed., rev. New York: John Wiley and Sons, 1912.

Winchester, Simon. "Taking the World's Measure." *Civilization* 2 (November/December 1995): 56–59.

Wozencraft, Jennifer M., and W. Jeff Lillycrop. "SHOALS Airborne Coastal Mapping: Past, Present, and Future." *Journal of Coastal Research*, special issue 38 (Fall 2003): 207–15.

Wright, John Kirtland. *Geography in the Making: The American Geographical Society, 1851–1951*. New York: American Geographical Society, 1952.

Zahl, Paul A. "The Giant Tides of Fundy." *National Geographic Magazine* 112 (August 1957): 153–92.

Zerger, A., and others. "Riding the Storm: A Comparison of Uncertainty Modelling Techniques for Storm Surge Risk Management." *Applied Geography* 22 (2002): 307–30.

Zhang, Keqi, and Dean Whitman. "Comparison of Three Algorithms for Filtering Airborne Lidar Data." *Photogrammetric Engineering and Remote Sensing* 71 (2005): 313–24.

Zhang, Keqi, Chengyou Xiao, and Stephen Leatherman. *Storm Surge Simulation for Lake Okeechobee*. Miami, Fla.: Florida International University, International Hurricane Research Center, 2006.

INDEX

Page numbers in italics refer to figures.